Tropical
FOREST BIOMES

Barbara A. Holzman

Greenwood Guides to Biomes of the World

Susan L. Woodward, General Editor

GREENWOOD PRESS

Westport, Connecticut • London

Library of Congress Cataloging-in-Publication Data

Holzman, Barbara A.
    Tropical forest biomes / Barbara A. Holzman.
        p. cm. — (Greenwood guides to biomes of the world)
    Includes bibliographical references and index.
    ISBN 978-0-313-33840-3 (set : alk. paper) — ISBN 978-0-313-
33998-1 (vol. : alk. paper)
    1. Rain forest ecology. 2. Rain forests. I. Title.
QH541.5.R27H65 2008
577.34—dc22   2008027506

British Library Cataloguing in Publication Data is available.

Library of Congress Catalog Card Number: 2008027506
ISBN: 978-0-313-33998-1 (vol.)
        978-0-313-33840-3 (set)

First published in 2008

Greenwood Press, 88 Post Road West, Westport, CT 06881
An imprint of Greenwood Publishing Group, Inc.
www.greenwood.com

Printed in the United States of America

The paper used in this book complies with the
Permanent Paper Standard issued by the National
Information Standards Organization (Z39.48–1984).

10   9   8   7   6   5   4   3   2   1

To Dave, Jacob, and Laurasia
for their continuous and unwavering
love and support

Contents

How to Use This Book

The book is arranged with a general introduction to tropical forest biomes and chapters on the Tropical Rainforest Biome and Regional Expressions of the Tropical Rainforest Biome, and the Tropical Seasonal Forest Biome and Regional Expressions of the Tropical Seasonal Forest Biome. The biome chapters begin with a general overview at a global scale and proceed to regional descriptions organized by the continents on which they appear. Each chapter and each regional description can more or less stand on its own, but the reader will find it instructive to investigate the introductory chapter and the introductory sections in the later chapters. More in-depth coverage of topics perhaps not so thoroughly developed in the regional discussions usually appears in the introductions.

The use of Latin or scientific names for species has been kept to a minimum in the text. However, the scientific name of each plant or animal for which a common name is given in a chapter appears in an appendix to that chapter. A glossary at the end of the book gives definitions of selected terms used throughout the volume. The bibliography lists the works consulted by the author and is arranged by biome and the regional expressions of that biome.

All biomes overlap to some degree with others, so you may wish to refer to other books among Greenwood Guides to the Biomes of the World. The volume entitled *Introduction to Biomes* presents simplified descriptions of all the major biomes. It also discusses the major concepts that inform scientists in their study and understanding of biomes and describes and explains, at a global scale, the environmental factors and processes that serve to differentiate the world's biomes.

The Use of Scientific Names

Good reasons exist for knowing the scientific or Latin names of organisms, even if at first they seem strange and cumbersome. Scientific names are agreed on by international committees and, with few exceptions, are used throughout the world. So everyone knows exactly which species or group of species everyone else is talking about. This is not true for common names, which vary from place to place and language to language. Another problem with common names is that in many instances European colonists saw resemblances between new species they encountered in the Americas or elsewhere and those familiar to them at home. So they gave the foreign plant or animal the same name as the Old World species. The common American Robin is a "robin" because it has a red breast like the English or European Robin and not because the two are closely related. In fact, if one checks the scientific names, one finds that the American Robin is *Turdus migratorius* and the English Robin is *Erithacus rubecula*. And they have not merely been put into different genera (*Turdus* versus *Erithacus*) by taxonomists, but into different families. The American Robin is a thrush (family Turdidae) and the English Robin is an Old World flycatcher (family Muscicapidae). Sometimes that matters. Comparing the two birds is really comparing apples to oranges. They are different creatures, a fact masked by their common names.

Scientific names can be secret treasures when it comes to unraveling the puzzles of species distributions. The more different two species are in their taxonomic relationships the farther apart in time they are from a common ancestor. So two species placed in the same genus are somewhat like two brothers having the same father—they are closely related and of the same generation. Two genera in the same family might be thought of as two cousins—they have the same grandfather, but different fathers. Their common ancestral roots are separated farther by time. The important thing in the study of biomes is that distance measured by time often means distance measured by separation in space as well. It is widely held that new species come about when a population becomes isolated in one way or another from the rest of its kind and adapts to a different environment. The scientific classification into genera, families, orders, and so forth reflects how long ago a population went its separate way in an evolutionary sense and usually points to some past environmental changes that created barriers to the exchange of genes among all members of a species. It hints at the movements of species and both ancient and recent connections or barriers. So if you find a two species in the same genus or two genera in the same family that occur on different continents today, this tells you that their "fathers" or "grandfathers" not so long ago lived in close contact, either because the continents were connected by suitable habitat or because some members of the ancestral group were able to overcome a barrier and settle in a new location. The greater the degree of taxonomic separation (for example, different families existing in different geographic areas) the longer the time back to a common ancestor and the longer ago the physical separation of the species. Evolutionary history and Earth history are hidden in a name. Thus, taxonomic classification can be important.

Most readers, of course, won't want or need to consider the deep past. So, as much as possible, Latin names for species do not appear in the text. Only when a common English language name is not available, as often is true for plants and animals from other parts of the world, is the scientific name provided. The names of families and, sometimes, orders appear because they are such strong indicators of long isolation and separate evolution. Scientific names do appear in chapter appendixes. Anyone looking for more information on a particular type of organism is cautioned to use the Latin name in your literature or Internet search to ensure that you are dealing with the correct plant or animal. Anyone comparing the plants and animals of two different biomes or of two different regional expressions of the same biome should likewise consult the list of scientific names to be sure a "robin" in one place is the same as a "robin" in another.

Acknowledgments

In preparation of this volume, I have received assistance from numerous colleagues. The Organization for Tropical Studies provided assistance on my visits to their Costa Rican Biological Stations. Field Station Directors Zak Zahawi at Las Cruces, Eugenio Gonzales at Palo Verde, and Deedra McClearn at La Selva and their staff were most helpful. Dori Dick, Miguel Fernandez, and Jeff Mitchell contributed initial text on conservation of tropical forests. Dr. Robert Drewes of California Academy of Sciences was generous with his photos and time. San Francisco State University provided assistance in terms of a one-semester sabbatical. Dean Joel Kassiola and the College of Behavioral and Social Sciences and the Department of Geography provided support for travel. Lastly, thank you to Susan Woodward, who provided much patience, support, and editing of this volume.

1

Introduction to Tropical Forest Biomes

In the region directly north and south of the Equator called the Tropics lies the most unique and diverse lands. These lands are home to hundreds of thousands, or perhaps millions of species, and regulate the Earth's air, water, weather, and energy cycles. These lands and their flora and fauna are essential to life today. On these lands are the Tropical Forests of the world.

Tropical Forest Biomes hold the richest biodiversity of all other terrestrial biomes. Two world biomes are included in the Tropical Forest designation, the Tropical Broadleaf Evergreen Forest, also called the Tropical Rainforest or the Equatorial Rainforest Biome, and the Tropical Seasonal Forest also known as the Tropical Deciduous Forest or the Tropical Monsoonal Forest Biome. Both biomes are located between the Tropic of Cancer at 23° N latitude and the Tropical of Capricorn at 23° S latitude. The zone between these two latitudes is called "the Tropics." "The Equatorial Zone" is within the Tropics, along both sides of the Equator. Tropical rainforests occur mostly within the Equatorial Zone. Tropical Seasonal Forests are located north and south of the equatorial zone. High temperatures, caused by intensive solar radiation and high rainfall amounts are universal in the tropics. The Tropical Rainforest experiences these throughout the year, while the Tropical Seasonal Forest experiences seasonal shifts in temperature, and drastic seasonal shifts in rainfall. The plants, animals, and other organisms that inhabit these biomes have evolved unique and varied adaptations to thrive in these environments.

This volume provides information on both Tropical Forest Biomes: the Tropical Rainforest and the Tropical Seasonal Forest. Each biome is described, beginning with a general global overview where the following factors are discussed:

- Geographic location
- Formation and origin of the biome
- General climatic conditions of the biome
- Major soil-forming processes and types of soil in the biome
- Common structure and characteristics of tropical forest vegetation
- Common adaptations and types of animals living in the biome
- Current condition of the biome and efforts toward conservation

Following the general overview, regional expressions of the biome are discussed; details of location, climatic influences, soils, specific plant and animal species, and their adaptations, along with a description of cultural influences and regional conservation, are presented.

The Tropical Forests

The Tropical Forests of the world have their geographic or latitudinal location in common. Tropical Forests are located between the Tropic of Cancer at 23° N and the Tropic of Capricorn at 23° S of the Equator. The Tropical Rainforest Biome occurs closer to the Equator where day length, temperature, and rainfall are consistent throughout the year. The Tropical Seasonal Forest Biome is present away from the equatorial zone, toward the tropical margins. In a few areas, these tropical forests stretch beyond the tropics, due to specific oceanic or climatic influences. Tropical rainforests make up about 86 percent of tropical forest; the remaining 14 percent is composed of Tropical Seasonal Forest. These Tropical Forest Biomes occur in three regions: Central and South America; West, Central, and Interior Africa and Madagascar; and in the Asian Pacific, Southeast Asia, New Guinea, and northeastern Australia.

Tropical forests cover approximately 7 percent of the Earth or close to 5 million acres (2.1 million ha). Of that 7 percent, approximately 45 percent is located in the Americas, while the Asian Pacific and African regions have roughly 25 and 30 percent, respectively (see Figure 1.1).

Tropical Forest Biomes contain the highest biodiversity of all other terrestrial biomes. These biomes cover a small area compared with other biomes, but play a crucial role in the atmospheric, climatic, and ecological systems of the world.

Vegetation

Tropical rainforests are broadleaved evergreen forests found at elevations below 3,300 ft (1,000 m); temperatures are warm year-round and precipitation is high.

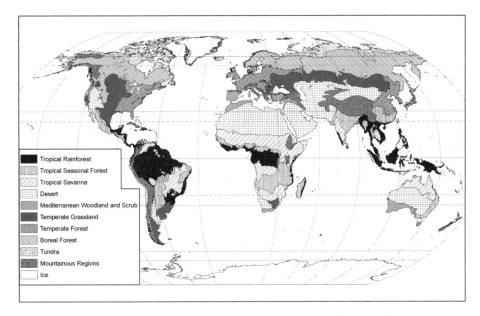

Figure 1.1 Tropical Forest Biomes of the world. *(Map by Bernd Kuennecke.)*

Tropical seasonal forests are broadleaved deciduous forests and evergreen dry forests found in regions that are both warm year-round and strongly water limited with several months of drought.

While both biomes have many things in common, they can be easily distinguished by climate, soils, vegetative structure, and types of plants and animals. Tropical plant communities form a graded series from wet to dry. Yet this change is so gradual that no clear boundary exists between the two biomes.

Some characteristics of the Tropical Rainforest Biome include more than 100 ft (30 m) tall emergent trees, with buttressed bases, understory plants with large evergreen leaves with rain-adapted "drip tips," and hundreds of species of epiphytes and woody vines that hang on trees throughout the forest canopy. On the margins of the tropics, drier or xeric plant communities such as thorn forest have short trees less than 15 ft (5 m) tall and lacking buttresses. They have relatively small compound leaves and trunks armed with thorns. Trees of the Tropical Seasonal Forest Biome fall in the middle of the spectrum, with broadleaf trees and vines that lose their leaves at the start of the dry season and proceed to flower and fruit.

Tropical rainforests are densely populated by trees and others plants that form many vegetative layers within the forest. The uppermost layers form the canopy. As you move down through the canopy, each lower layer is influenced by the upper layers. The highest canopy layer consists of scattered, very tall trees called emergents that encounter extreme solar input (heat and light) as well as rainfall and wind. The next canopy layer, below the emergent trees, is densely forested, allowing little light to penetrate to the lower forest layers. Lower canopy layers house progressively shorter trees. There may be as many as five or more layers of trees, shrubs, or other growthforms within a tropical forest. A dense cover of woody vines, epiphytes, and parasitic and carnivorous plants are found throughout the canopy. Lower canopy layer vegetation typically has wide leaves to capture the limited and filtered sunlight that escapes the upper canopy. Little plant life grows on the forest floor, as the upper layers tend to make use of most all of the sunlight and little reaches the ground. In the Tropical Seasonal Forest, the canopy is less complex. Because many of the trees lose their leaves during the dry season, a dense understory often develops.

Species are abundant in the Tropical Forest Biomes; however, in many cases their populations are small. Many species within tropical forests have restricted distribution areas. There may be as many as 100–300 different trees within 2.4 ac (1 ha) of tropical rainforest, but few will be found in other parts of the forest. Significantly more species than genera are found in the Tropical Forests; in other words, the species to genera ratio is high and many congeners—species in the same genus—are found. Nonetheless, the numbers of both species and genera far exceed the numbers in other biomes. Several congeners may be found in one part of the forest, while a different set is found in another area. The floral composition is much richer and more complex in the Tropical Rainforest than the Tropical Seasonal Forest. Both are richer and more complex than other biomes.

Climate

Equal day length, constant heat and moisture, and the effects of large water bodies surrounding most of the tropics provide the Tropical Forest Biomes with temperatures that are constant throughout the year. In general, the Tropical Rainforest Biome's average annual temperature ranges from 79°–81° F (26°–27° C). Daily temperatures are more varied than annual temperatures, shifting as much as 8° F (4.5° C) in some areas, depending on cloud cover and rainfall. As you leave the equatorial zone, the average annual temperature begins to vary slightly. In the Tropical Seasonal Forests, temperatures are highest when the sun is not directly overhead. Seasonal temperatures range between 68°–86° F (20°–30° C). Daily temperatures in the dry season remain around 82° F (28° C), while average daily temperatures in the wetter season are around 78° F (26° C) due to cloud cover.

Tropical Forests need heat and abundant moisture. The presence of frost is a limiting factor. Frost limits plant growth and can be lethal for most tropical vegetation. Frost also limits the expansion of the Tropical Forest Biomes beyond 23° N and 23° S latitude. These low latitude climatic regions receive 100–180 in (2,500–4,500 mm) of precipitation annually. The rain falls throughout the year with a seasonal decrease in some areas. In the rainforest, the dry season is not truly dry. During this time, the rain falls intermittently, with an occasional week or two with no rain. Precipitation amounts decrease as you move away from the Equator.

In monsoonal and seasonal forests, there is a true dry season, usually for four to seven months, but that is compensated for by abundant precipitation throughout the rest of the year.

In Tropical Forests, day length (photoperiod) and the angle at which sunlight strikes the Earth's surface varies little throughout the year. Slight shifts can create the seasonal weather patterns that typically occur in the Tropical Seasonal Forest Biome. In all parts of these biomes, gradual changes in climate patterns occur across large distances.

Global Circulation

Two major global circulation systems affect the Tropical Forests of the world. They are the Intertropical Convergence Zone (ITCZ) and the Trade Winds. These two systems regulate the climate and weather patterns in the tropics. With abundant and direct sunlight, the equatorial zone is an intensive source of heat and moisture. As airmasses are intensely heated at the Equator, they rise into the atmosphere. This rising air creates a pressure gradient, with low pressure at the Equator and higher pressure at the mid-latitudes. The pressure gradient causes a shifting of the warmer airmasses toward the pole in the upper atmosphere, and a shifting of colder airmasses from mid-latitudes toward the Equator in the lower atmosphere. This movement of air toward the Equator occurs in the Northern and Southern hemispheres and initiates air movement and the general circulation patterns around the globe (see Figure 1.2). As these airmasses flow together at the Equator, a zone of unstable air is created. This is called the ITCZ. The ITCZ occurs between about 5° N and 5° S of the Equator, but can extend farther in certain regions. The ITCZ moves north or south depending on the season and amount of solar energy received. There is often a difference in the ITCZ boundary on land and over the oceans. Above the oceans, unstable air (generating heavy rains) and unpredictable winds, or in some cases no wind at all, are typical within the ITCZ. This zone at sea was known as the "doldrums" by earlier mariners.

The movement of air from the mid-latitudes toward the Equator creates a general wind circulation pattern north and south of the ITCZ. The typically steady flows of air known as the Trade Winds facilitated the movement of trade from Europe to the Americas and back again. This trade wind circulation is most prominent over the Atlantic and Pacific oceans and least in the monsoon-controlled regions surrounding the Indian Ocean. Trade wind circulation is an easterly flow in both hemispheres—that is, the winds travel from the east toward the west. The winds tend to converge toward the Equator into the ITCZ. The Trade Winds are capable of bringing large amounts of rainfall, especially on the east coast of tropical countries, such as in the Caribbean and in Central America.

On the continents, the ITCZ shifts poleward from the Equator into the subtropics with the solstices. These seasonal shifts can bring strong winds and heavy rains into certain areas north and south of the tropics. Differences in temperature between land and sea during changing seasons causes shifts in wind patterns. These

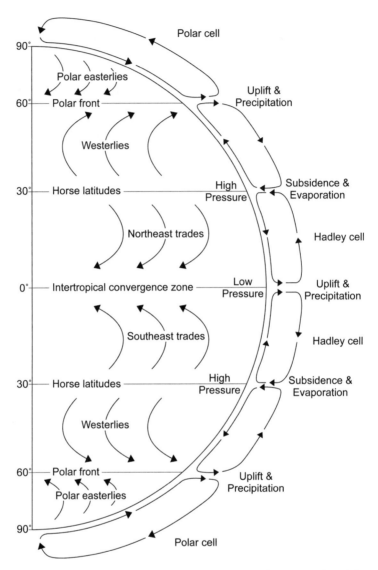

Figure 1.2 General global circulation and wind patterns in the tropics. *(Illustration by Jeff Dixon.)*

seasonal shifts in winds are called monsoons. Large centers of population and intensive agriculture within these areas depend on these monsoonal rains. These high-intensity airmasses with high water content can bring torrential rains over short periods. As much as 20–30 in (500–700 mm) of *daily* rainfall have been recorded in sites in Africa and India. Monsoons are discussed further in Chapter 4.

In the continental or interior tropics, trade wind circulation is less pronounced and differences in heat energy exchange between the forest and atmosphere are more influential. Interior rainforests typically experience convective precipitation that is caused by warm, humid air (from evaporation processes within the forest) at the surface rapidly rising, cooling aloft. Similarly, as air rises up mountains it cools and the water vapor in the air condenses. Since cool air can hold less water, when the airmass becomes saturated rain falls. This typically happens in the late afternoon and can bring torrential rains into and through the night. Hurricanes, cyclones, or typhoons are low-pressure systems that develop in warm tropical waters. They build up speed and momentum at sea and when they reach land rainfall can be extremely intense. The accompanying strong winds can cause widespread flooding and destruction. Tropical cyclones are concentrated in the Pacific and Indian oceans affecting Madagascar, Southeast Asia, and Australia. Atlantic cyclones occur in the Caribbean and Central America.

The El Niño-Southern Oscillation (ENSO) is a cyclical, global ocean-atmospheric phenomenon that affects the tropical regions of the world. El Niño and its counterpart La Niña influence temperature changes in the surface waters of the tropical Pacific Ocean, with a significant effect on the weather patterns, especially in the Southern Hemisphere. During an El Niño event, the warm waters of the central equatorial Pacific are pushed eastward to the normally cold water coasts of South America. This warm water with accompanying warm air above provides abundant moisture to some areas that are typically dry. In contrast, the easterly movement of warm water leaves the western Pacific without its normally abundant moisture. El Niño causes weather patterns that can produce excessive rain in some places and extreme drought in others. It has been linked to flooding in the tropical dry forests along the Pacific Coast of Central and South America and destructive drought and fires in the tropical forests of Asia and Australia. Tropical forests in Asia experience extreme drought during severe El Niño events, leaving them vulnerable to destructive fires that destroy entire forest stands. Global climate change is thought to have increased the frequency of El Niño events; however, research is ongoing to assess the true impacts of global warming on El Niño, as well as other climatic phenomenon.

Soils

Soils within the tropical zone are the product of high heat and moisture over millions of years. They are primarily highly weathered ancient soils with high acidity and low nutrient and organic matter content. They tend to be red or yellow in color. Other younger soils do occur within the tropics, and they tend to be more fertile and have a higher nutrient and mineral content than those developed on ancient surfaces. They are brown or black in color due to the higher levels of organic matter in the soil and their volcanic origins. Soil types are discussed in later chapters.

Evolutionary Processes

The fauna as well as flora of Tropical Forest Biomes are highly diverse. The length of time these regions have remained isolated with abundant heat and moisture have provided ample opportunity for animals to develop unique niches and strategies for survival. Continuous plant diversification through evolution leads to increasingly diverse habitats and abundant opportunities for the diversification of animal species. These millions of years of isolation with little floral or faunal exchange have created opportunities for speciation through adaptive radiation, the evolution of many species from a single ancestral species. This often happens in concert with isolation.

Another evolutionary phenomenon displayed in the tropical rainforests of the world is convergent evolution. Convergent evolution is the development of two or more species that resemble each other or share similar behaviors but are totally unrelated. Hummingbirds of the Neotropics and sunbirds in Africa and the

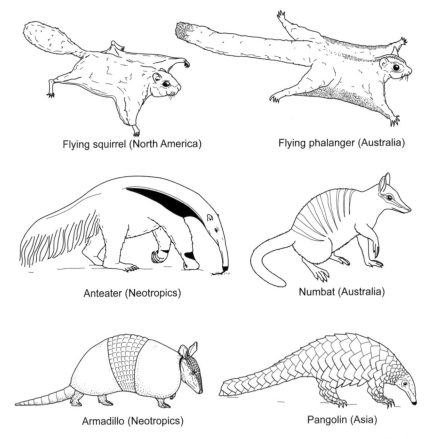

Flying squirrel (North America)   Flying phalanger (Australia)

Anteater (Neotropics)   Numbat (Australia)

Armadillo (Neotropics)   Pangolin (Asia)

Figure 1.3 Examples of convergent evolution. Although the animals shown look similar, they are unrelated. *(Illustration by Jeff Dixon.)*

Asian-Pacific rainforests look similar and fill similar niches. The Neotropical agouti and Asian deer mouse are also similar in appearance. Toucans and hornbills show remarkable similarities. Yet none of these animals are related. Figure 1.3 provides some additional examples of this phenomenon.

Adaptations for a Tropical Life

Plants and animals in the Tropical Forest Biomes, with millions of years of isolation and long periods of climatic stability, have evolved interesting strategies to respond to the environment. Adaptation is the way in which a plant or animal fits into its surroundings. The better adapted an organism is, the more likely it will survive. Adaptations are influenced by myriad stresses and demands encountered by the organism. Temperature, light, shade, water, food, shelter, as well as predation, competition, and reproduction are all opportunities for adaptations. Evolutionary adaptations have allowed tropical plants to flourish amid these many stresses. Organisms within the tropical forests have evolved to maximize their probability of survival and reproduction. As the world within the tropics has changed, so have the organisms that inhabit it. When an organism is no longer able to adapt to its changing environment, its chances of long-term survival are limited.

Stresses on tropical forest vegetation include high sun intensity; high rainfall; unseasonable and episodic droughts; high frequency of tropical storms, hurricanes, or typhoons; lightning strikes; severe wind events; and susceptibility to fire. Nutrient poor soils and supersaturated soils lead to the leaching of minerals and organic matter, as well as soil erosion. Tropical rainforests typically have no opportunity for seasonal recovery from these saturated conditions. Tropical Seasonal Forests can suffer from extreme events and can be stressed by either an abundance or total lack of moisture.

Adaptations evident in plants include variations in physical characteristics such as tree architecture, root structures, bark texture and thickness, variable leaf shape and size, seed shape and size, as well as pollination and germination strategies. Physiological strategies such as differences in photosynthetic rates, uptake of nutrients and wood density, as well as chemical strategies that discourage predation are all types of adaptation found in tropical plants. Many of these adaptations are described later in this volume.

The faunal composition of the Tropical Forest Biomes is similar within their tropical regions. Larger mammals and birds often migrate between the two biomes. Animals have developed unique evolutionary traits within each regional expression of the biome—strategies to travel through the forest, to take advantage of the many and varied food resources available, to attract mates, and to discourage predation. These adaptations are evident in vertebrates and invertebrates alike. In Australasia, there has been a full diversification of marsupial mammals in the region. These include adaptations to life in the trees or on the forest floor. Specific animals and their adaptations to their regional locations are discussed in Chapters 3 and 5 under each section on Regional Expressions of the Biome.

The Tropical Forest Biomes are the most ancient, diverse, and ecologically complex biomes on the planet and support more than half of the Earth's species. At least 3 million species are known to inhabit the tropical forests of the world. This number could be 10 or more times greater, because not all tropical species have been scientifically described. New species within these forests continue to be discovered.

Table 1.1 Comparison of Tropical Rainforest and Seasonal Forest Biomes

	TROPICAL RAINFOREST	TROPICAL SEASONAL FOREST
Location	Along the Equator, between 10° N and S latitude	North and south of the tropical rainforest (10°–23° N and S latitudes)
Temperature controls	Tropical latitudes, constant solar insolation	Tropical latitudes, shifting solar insolation
Temperature patterns	Little to no variation	Minimal seasonal variation
Precipitation controls	ITCZ, convective showers	Shifting ITCZ and Trade Winds
Precipitation totals	High	Seasonal drought
Seasonality	No	Yes
Climate type	Tropical wet	Tropical wet and dry / Tropical seasonal wet
Dominant growth forms	Broadleaf evergreen trees, vines, epiphytes	Broadleaf deciduous trees, vines, epiphytes
Major soil orders	Oxisols, ultisols, inceptisols, and entisols	Oxisols, ultisols, inceptisols, and entisols
Soil characteristics	Low fertility, low nutrients, highly weathered, acidic, red to yellow in color	Higher fertility, some nutrients, weathered, red to brown in color
Biodiversity	Highest in world	Very high
Age	Ancient: Tertiary to Mid-Cretaceous origin	Ancient: Tertiary to Mid-Cretaceous origin
Current status	Highly threatened with total extinction due to severe deforestation, mining, changes in land use that affect climate patterns within the forest, high development and population pressures, and climate change	Highly threatened/endangered due to severe deforestation, mining, changes in land use affecting climate patterns within the forest, high development and population pressures, and climate change

Note: ITCZ = Intertropical Convergence Zone.

Similarities exist between the two Tropical Forest Biomes, as do distinct differences. Table 1.1 compares the Tropical Rainforest Biome with the Tropical Seasonal Forest Biome describing these similarities and differences.

Tropical Forests contribute far more to the world than just sustaining biodiversity. They provide habitat and homes to indigenous people, a multitude of natural products such as food, building materials, and medicines, as well as a number of ecosystem services including soil stabilization and flood prevention. More than 20 percent of the world's oxygen is produced by the Amazon Rainforest of South America. One-fifth of the world's freshwater is found in the Amazon Basin. Tropical rainforests are critical in maintaining the Earth's limited supply of fresh water. These forests are also important sources of carbon storage and play key roles in both regional and global climates.

Humans and the Tropical Forest Biomes

Deforestation can have devastating effects on the Tropical Forest Biomes of the world. Deforestation affects biodiversity through habitat destruction, fragmentation of formally contiguous habitat, and increased edge effects. Species loss is much greater in the tropics than in any other biome. Continuous and increased occupation of forested land has created permanent changes to the forest landscape. The extensive deforestation of tropical forests is a source of greenhouse gases (carbon dioxide and methane) through forest burning. As forests are cleared, less carbon dioxide is absorbed and stored through photosynthesis. The addition of human-caused inputs of carbon dioxide as well as other greenhouse gases into the atmosphere has significantly increased since the start of the Industrial Revolution. The effects of greenhouse gases and global warming could lead to warmer and wetter tropical areas. However, the expansive losses of forests have changed these effects. Increased regional albedo (Earth's reflectivity) due to deforestation alters rainfall patterns in the tropical and temperate latitudes affecting critical climatic processes. While tropical areas of Asia and Central Africa may see warmer and wetter conditions, tropical areas in South America and West Africa may find the environment warmer, but drier. Other land use activities destructive to the tropical forests include gold mining, mineral extraction, conversion for small and large-scale agriculture, and livestock ranching. Road building and oil drilling also contribute to the loss of species and tropical forest integrity. Many emerging countries are home to the Tropical Forest Biomes of the world; increasing population pressure and economic development amplify the stresses put upon tropical forests.

Tropical Forest Biomes are unique environments. They house the richest biodiversity in the world. These biomes provide humans with food, shelter, medicines, and other natural resources. Their life-sustaining processes are crucial in climate regulation, oxygen production, and water cycling. The fate of these tropical biomes will affect the fate of the rest of the world.

Further Readings

Books

Aubert De La Rue, E., Francois Bourliere, and Jean-Paul Harroy. 1957. *The Tropics*. New York: Alfred A. Knopf.

Lieth, H., and M. J. A. Wegener, eds. 1989. *Tropical Rain Forests Ecosystems: Biogeographical and Ecological Studies*. New York: Elsevier.

Mabberley, D. J. 1991. *Tropical Rain Forest Ecology*. Oxford: Chapman and Hall.

Miller, K., and Laura Tangley. 1991. *Trees of Life: Saving Tropical Forests and Their Biological Wealth*. Boston: Beacon Press.

Richards, P. W. 1996. *The Tropical Rain Forest: An Ecological Study*. Cambridge: Cambridge University Press.

Internet Source

Cool Planet. 2002. *Tropical Rain Forest*. Oxfam. http://www.oxfam.org.uk/coolplanet/ontheline/explore/nature/trfindex.htm.

2

The Tropical Broadleaf
Evergreen Forest Biome

The Tropical Broadleaf Evergreen Forest Biome is also known as the Tropical Rainforest or the Equatorial Forest Biome. This biome is a type of broadleaf evergreen forest found near the Equator (see Figure 2.1). Although tropical forests cover merely 7 percent of the Earth's surface, terrestrial biodiversity is highest in the Tropical Rainforest Biome.

Geographic Location

The Tropical Rainforest Biome is located in the equatorial latitudes (10° N through 10° S latitude) on both hemispheres below 3,300 ft (1,000 m) elevation. Tropical rainforests can extend beyond these latitudes in areas with high moisture conditions. Three major, separate formations or regional expressions of the Tropical Rainforest Biome (see Figure 2.2) occur:

- Neotropical Rainforest: Central and South America, Caribbean Islands
- African Rainforest: West and Central Africa and their coastal islands, the Congo Basin, and eastern Madagascar
- Asian-Pacific Rainforest: west coast of India, Southeast Asia, Indonesia, New Guinea, and northeastern Australia

Similar niches—multidimensional resource space—are found throughout these three areas; however, the species, as well as higher taxonomic groups (genera and

Figure 2.1 Tropical Rainforest in Guatemala. Note the presence of emergent trees. *(Photo by author.)*

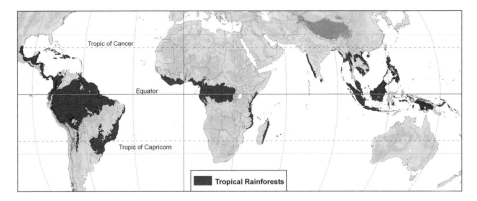

Figure 2.2 World distribution of Tropical Rainforest Biome. *(Map by Bernd Kuennecke.)*

families) are distinct in each. The species in the Tropical Rainforest Biome are vastly different from those of the temperate forests. Species diversity is highest in the extensive Neotropical Rainforest, particularly in the Amazon; second in the highly fragmented Asian-Pacific region; and lowest in Africa.

Tropical Rainforests

Tropical rainforests hold the largest amount of biomass of any terrestrial habitat. Apart from major climatic events and regional climatic fluctuations, tropical rainforests have remained fairly stable and intact for hundreds of millions of years. They are the planet's major genetic reserve, with plants, mammals, birds, reptiles, amphibians, fish, invertebrates, and microorganisms constantly evolving to fill in niches within this rich environment.

With this constantly evolving system, it might be expected that these areas, once isolated from each other, would have few similarities. Tropical rainforests throughout the world exhibit wonderful differences, but distinctive floristic and faunal similarities still exist throughout all of the tropical rainforests of the world.

These similarities in flora and fauna can be partly explained by organisms traveling around the world for millions of years and finding a similar, hospitable environment in another tropical forest. This process of movement across vast gaps made by unsuitable habitat followed by colonization is called long-distance dispersal. The age of these forests and the time elapsed to accommodate long-distance dispersal makes these events probable. Most tropical rainforests evolved within the regime of constant rainfall and persistent heat, still characteristic of their locations, so similar environments would have been accommodating for newly arriving species. While this helps to understand some of the commonalities between the rainforest regions, their unique floristic similarities beg further review.

Formation and Origin of the Tropical Rainforest Biome

Understanding the Tropical Rainforest and the species it holds begins with understanding its origin. Although long-distance dispersal can explain some similarities among the tropical forests of the world, many of the similarities and evolutionary relationships between tropical forests are a factor of their common geologic history. These ancient landscapes originated while in close contact with each other. The continents that currently house tropical forests started out as a large landmass near the South Pole about 300 million years ago (mya), in the early Paleozoic.

For hundreds of millions of years, this landmass slowly moved toward the Equator through the geologic process of plate tectonics. About 200 mya, this land mass met up with a more northern and equatorial landmass and formed a supercontinent called Pangaea. Pangaea means "all land." During this time, plants and animals were able to spread throughout the supercontinent, only limited by mountain ranges and subtropical deserts. Fragments of the ancient bedrock or shield that formed this supercontinent remain on many continents, bearing witness to their ancient connections. Pangaea was centered along the Equator, an area of high sun, heat, and moisture. This tropical environment provided an ideal setting for the evolution of the Tropical Rainforest Biome of today.

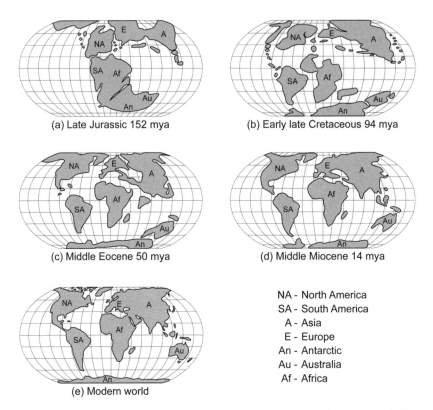

(a) Late Jurassic 152 mya

(b) Early late Cretaceous 94 mya

(c) Middle Eocene 50 mya

(d) Middle Miocene 14 mya

(e) Modern world

NA - North America
SA - South America
A - Asia
E - Europe
An - Antarctic
Au - Australia
Af - Africa

Figure 2.3 Plate tectonic movement as it relates to Gondwana. *(Illustration by Jeff Dixon.)*

The large landmass that once contained the tropical rainforests of South America, Africa, Madagascar, Southeast Asia, India, Australia, and New Guinea was part of the southern half of the supercontinent Pangaea. Pangaea remained centered along the Equator from the Triassic to the mid-Jurassic Period (240–160 mya). During this time, the continental blocks that form the core of Southeast Asia are thought to have separated from the supercontinent. As tectonic plates continued to shift, Pangaea began to break apart, and northern and southern landmasses were formed. The southern landmass containing Africa, India, Madagascar, South America, Australia, and Antarctica was called Gondwana (Gondwanaland). The more northern landmass containing North America and Eurasia was called Laurasia (see Figure 2.3). Laurasia moved away from the Equator to the northeast in a counter-clockwise motion. Gondwana remained along the Equator, experiencing high sun, high heat, and abundant rainfall near the coasts and continued to provide ideal conditions for the evolution of the Tropical Rainforest. The time and relative tectonic stability within Gondwana allowed the area to develop intricate and complex ecosystems.

Gondwana began to slowly separate; South America, Antarctica, Australia, and New Guinea began to move south and eastward due to the formation of the mid-oceanic ridge that created the Atlantic Ocean. In the early Cretaceous Period (around 120 mya), Madagascar and India separated from the Gondwanan landmass and began to move east and slightly north. Africa remained isolated for a long period during the late Cretaceous and early Tertiary Periods, developing a large endemic group of mammals. By about 50 mya, the continents began to resemble today's configuration. Each of the Tropical Rainforest Biome regions of the world carries a bit of that geologic history. Fossils of primitive flowering plants and early fauna, along with evolutionary relationships discovered between current species, provide evidence of their connected past. Even today, the similarities between these forests reveal these earliest connections.

At the time of Pangaea, the land was dominated by gymnosperms (conifers, ginkgos, and cycads) and ferns, which were adapted to a drier, more continental climate. Gymnosperms were the first seed plants; they have their ovules and seeds exposed on the plant surface; that is, they have naked seeds. This is in contrast to the flowering plants, angiosperms, whose ovules are contained within the flower and their seeds within a fruit. Even before the Pangaean supercontinent was created, primitive gymnosperms were present and evolving on the earlier southern landmass as it moved toward the Equator. Soon the evolution of flowering plants began. Although questions about the origin and evolution of flowering plants remain, current evidence points to a great explosion of angiosperms during the Cretaceous Period (around 144–65 mya), between 45° N and 45° S latitude, although some may have been present prior to that time. Recently, fossils of primitive angiosperms found in China were dated to the mid-Jurassic Period (175 mya). The abundance of moisture, sunlight, and heat helped support this great evolution and diversification of angiosperms, and these flowering plants began to dominant the central part of the world. At this time, Africa and South America remained close to each other and were able to share species.

The tropical areas of Gondwana experienced abundant sunlight, rain, and constant warm temperatures for millions of years. During the period when gymnosperms and ferns were dominant, dinosaurs roamed the land. Early mammals were just beginning to emerge. With the extinction of the dinosaurs in the Jurassic and Cretaceous periods (200–65 mya), habitat and niches became available for an explosive diversification of mammals, which rapidly began to evolve and diversify. Marsupial mammals arose between 250–100 mya. Evolutionary changes in reproduction and regeneration in flowering plants occurred simultaneously with mammal evolution as plants provided food for the mammals and the mammals aided the pollination, seed dispersal, and germination of plants. At this time South America, Antarctica, Australia, and New Guinea were separating from Africa. They began to move south from equatorial Gondwana, carrying some of the more primitive mammals, as well as primitive vascular plants and gymnosperms, with them. The plate containing Australia and New Guinea broke away from South America and began to move in a northeastern direction. This island block remained isolated for millions of years, and ancient gymnosperms and marsupials were able to gain a competitive foothold and continue to evolve, filling similar niches that placental mammals would fill in Africa, Asia, and the Americas.

The closed canopy forests we see in today's Tropical Rainforest Biome became widespread around 65–50 mya, in the early Tertiary Period. Recently, a hypothesis has been put forward that suggests that the similarities we see in some taxonomic groups among regions are caused by widespread global warming trends in the more recent geologic past. During times of increased temperature and higher precipitation levels, tropical species were more widely distributed—even at latitudes poleward of the tropics, allowing for periodic interchanges of plants and animals among the continents. Evidence points to a continuous tropical forest belt during the Tertiary Period, spreading from Southern Europe through Central Africa into Madagascar, and through southern Asia and the Far East. This forest was also present on South America. This new idea increases our understanding of evolutionary relationships and current distributions.

As the continental plates separated, changing climate began to restrict the Tropical Rainforest Biome. When cooling occurred, rainforest species survived only in those areas that remained tropical in climate. During several epochs, tropical species were restricted, ending in their current distribution during the Pleistocene Epoch. Beginning about 1.8 mya, with the onset of glaciation, the northern latitudes became colder and drier environments with greater seasonality, no longer favorable for tropical species. More pollen and fossil analyses, along with evolutionary studies, are needed to confirm this hypothesis regarding tropical species origins and distributions.

The biodiversity of the three main centers of tropical rainforest today is largely a consequence of evolution along with past climatic events. The Neotropical rainforest has the greatest diversity. Its large extent may have allowed more species to endure through the cooling climate more successfully than their African counterparts. Isolation from other continents produced a diverse array of species adapted specifically to that environment. African rainforests contracted significantly from the effects of cooling and drier climates, leaving small refugia, which became centers of endemism still evident today. The Asian-Pacific rainforest was less affected by the changing climate, although changes in sea level led to wider dispersion of plants and animals.

Tropical rainforests continuously adapted to changes in climate and geologic events. Throughout the changing geology and climate of the past, cooler glacial periods restricted rainforests, possibly confining rainforest species to narrow refuges and small remnants near the Equator. During warming events, the rainforest expanded poleward beyond the tropical zone. Species in the world's rainforests evolved and modified, adapting to the changing environment. The paleoecological record provides evidence of past changes in the distribution of tropical rainforests and helps in understanding the similarities found among them. Today, geographically separated rainforests contain their own arrays of species as well as associations derived from ancestors that had become established over 65 mya. Although many of the species found in the rainforest today reveal this ancient history, the current structure and appearance of the rainforest may be very different from the rainforests of the past.

Climate

The key to the lush vegetation and high species diversity in the Tropical Rainforest Biome is the climate. Tropical rainforests experience warm temperatures throughout the year, averaging 79°–81° F (26°–27° C). Average annual precipitation is between 100–180 in (2,500–4,500 mm). Tropical areas closer to the tropical margins experience a decreased annual rainfall, with the exception of those areas subject to intense monsoonal activity such as in Southeast Asia. Individual areas on the west side of continents, such as Cameroon in Africa, and between Panama and Ecuador in South America, have recorded even higher rainfall, as much as 475 in (12,000 mm) at Mount Cameroon, and 400 in (10,000 mm) in Colombia.

As mentioned in Chapter 1, several major wind and climatic systems influence the Tropical Rainforest Biome. These include the Northeast and Southeast Trade Winds, the Intertropical Convergence Zone (ITCZ), tropical cyclones, and monsoons. Local convective precipitation contributes to rainfall totals within the rainforest. In the equatorial tropics, the high water content of air, in combination with the convergence of unstable airmasses, produces intensive vertical movement of moist air through convection. These are the areas of the world with the highest amounts of rainfall. The core regions are the Amazon Basin, the Congo Basin, and the Indomalayan archipelago.

Tropical cyclones also produce substantial rainfall in the tropics. These storms, in combination with summer monsoonal rains, are sufficient for the existence of rainforests away from the equatorial zone. Tropical cyclones, called typhoons in the Pacific and Indian Ocean influence vegetation on the eastern and southern parts of Asia, northeastern Australia, and Madagascar. Atlantic cyclones, called hurricanes in the Northern Hemisphere, primarily affect the Caribbean and Central American forests.

The climate of tropical rainforests is classified as a tropical moist climate, or Af using Koeppen's climate classification, where temperatures remain high and precipitation averages greater than 2.4 in (60 mm) per month (see Figure 2.4). Rainforests are located in and typically limited to the areas where precipitation is higher than evaporation, leading to a surplus of water. This creates a positive water balance. Tropical rainforests can extend beyond these areas when groundwater reservoirs are sufficient to overcome occasional drought events.

Taking a macroclimatic or large-scale view of this tropical biome, climate and weather play an important role in the distribution and physiology of vegetation and the internal forest cycles within these regions. Climate and weather are determined by incoming solar radiation (insolation), absorption, and reradiation (return of energy to the atmosphere) from the Earth's surface. Global atmospheric circulation and hydrologic cycles are also important components.

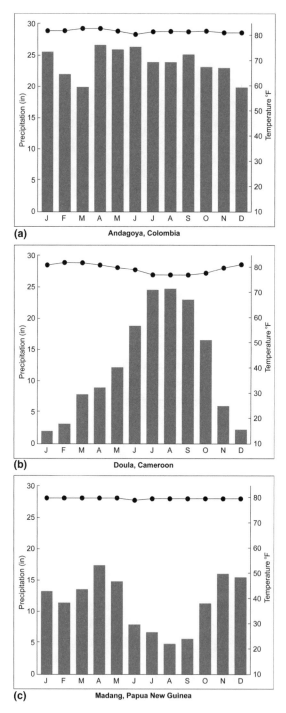

Figure 2.4 Climagraphs for Andagoya, Colombia, Douala, Cameroon, and Madang, PNG. In all three regions, high temperatures and high rainfall are constant throughout the year. *(Illustration by Jeff Dixon. Adapted from Richards 1996.)*

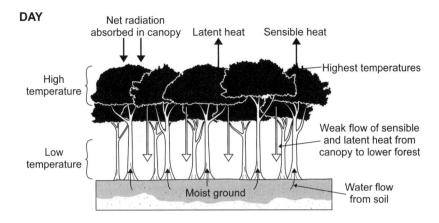

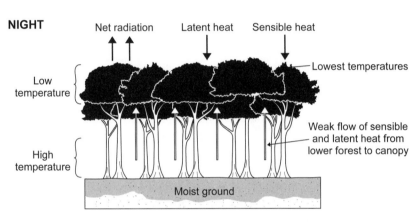

Figure 2.5 Energy dynamics within a tropical rainforest. Temperatures within the forest decrease with nightfall. *(Illustration by Jeff Dixon. Adapted from Lieth et al. 1989.)*

The Earth receives 99 percent of its energy from the sun. The tropical zone between 23° N and 23° S latitude receives the highest amount of this energy because of the Earth's revolution, rotation, and axial tilt. A constant input of the sun's energy, along with the Earth's axial tilt and revolution around the sun, results in minimal seasonal variation. The maximum day length ranges from 12 hours at the Equator to 13.5 hours at the Tropics of Cancer and Capricorn when the sun is directly overhead. Solar radiation is received directly or diffusely depending on cloud cover. The amount, course, and altitude of tropical clouds influence temperatures, but the growing season is always 365 days per year.

In the Tropical Rainforest, the transfer of energy from the sun to the forest primarily occurs in the forest canopy (see Figure 2.5). Less than 10 percent of the incoming solar radiation actually reaches the ground due to the density and height of canopy trees. The forest's own biological regulating mechanisms (photosynthesis, evaporation, transpiration, and so on) regulate energy transfer below the canopy, maintaining constant temperatures. The transfer of heat energy on the ground is minimal.

Very little of the solar radiation entering the forest is reradiated or returned to the atmosphere. The high water vapor content and high carbon dioxide levels, produced by plant respiration and decomposition within the forest, serve to absorb the outgoing heat energy, returning it to the tropical forest system. This leads to an almost uniform daily cycle in the radiation balance measured in the tropics. Insolation steadily increases from sunrise until noon, and gradually decreases until night fall. Patterns of annual (yearly) radiation (light and heat energy input) show high values during the equinoxes, with slightly decreased values during the rainiest season. In tropical areas farther from the Equator, a distinct seasonal change is noted. These fluctuations play a role in the distribution and physiology of some tropical plants at the tropical margins.

Heat Budget

Understanding the exchange of heat from the atmosphere to the canopy and forest floor is crucial in understanding tropical rainforests and their climate. These changes are more apparent daily than annually. During the day, insolation is primarily received by the tree canopy. Within the forest, energy is absorbed and released through evaporation and transpiration by plants. This evapotranspiration cycle causes variation in water vapor content within the forest. Despite these changes in energy and water vapor, the air temperature within the forest remains relatively constant at 79°–81° F (26°–27° C). Besides evapotranspiration, energy is released during the night through the condensation of water vapor. This release of energy is absorbed by the atmosphere, as well as the tree bases at ground level. These conditions regulate temperature and moisture and minimize drastic changes. In fact, the largest temperature variation and energy exchange occurs at the canopy level. These fluctuations in heat and energy transfer are more evident toward the subtropics, where pronounced wet and dry seasons occur.

Daily temperature changes influence tropical rainforest vegetation. Average daily temperatures show the greatest variation in open, treeless areas within the equatorial rainforest. Daily temperature variation is greater near the margins of the tropics (around 8° F [4.5° C]). These small diurnal temperature effects are also more pronounced in the central or interior parts of the rainforest, as coastal areas show minimal change. The minimum daily temperature occurs shortly before sunrise and temperature rises steadily reaching as much as 5°–9° F (3°–5° C) higher around noon, then decreases until nightfall and through the night. Cloud cover can add to or moderate these temperature changes.

Although the fluctuation of heat into and out of the forest can vary during the day, the annual cycle remains steady. Due to the steady input of light and heat, annual temperature cycles show little variation. Temperatures near the Equator change seasonally by as little as 2° F (1° C). The consistency in average monthly temperature is influenced by the consistent day length, as well as the large expanses of open water that occur in these tropical zones. These oceans moderate any sudden changes in temperature that may occur.

Rainfall Requirements of Tropical Rainforests

Seasonal cycles of rainfall are less evident near the Equator, and more pronounced toward the tropical margins, where maximum rainfall occurs after the sun reaches the equinoxes.

It is both the amount and continuity of rainfall that allows tropical rainforests to remain evergreen. Evergreen plants in the tropics can only exist where annual rainfall amounts, in combination with water storage capacity, are sufficient to maintain a positive water balance throughout the year. This means that the forest encounters minimal water stress. An annual rainfall of at least 100 in (2,500 mm) has been proposed as the minimum amount needed to maintain a tropical rainforest. However, this rainfall must be distributed evenly throughout the year. A negative water balance for more than a month has been shown to have a devastating effect on the tropical rainforest.

Like temperature, there is a diurnal cycle to rainfall within the Tropical Rainforest Biome. Rain typically occurs in the late afternoon when convection has reached its maximum. Rainfall can be very heavy at times, and in some areas can last until nightfall or well into the night. Inland regions and mountain slopes experience rain earlier in the afternoon compared with coastal areas, where rainfall typically occurs in the evenings.

Precipitation is cycled through small water cycles within forests, where precipitation evaporates through photosynthetic processes then condenses and is recovered as precipitation. These small water cycles have a transfer time of one to two days and account for nearly two-thirds of the water budget. The other way water is transferred is through the larger cycles that can encompass vast oceanic areas, such as rainfall transported through trade winds, monsoons, and cyclones. For example, the larger movements of water around the Atlantic Ocean feeds the African rainforest, and to a smaller extent the Amazon Basin.

These large and small water cycles contribute to the creation and maintenance of the Tropical Rainforest Biome. Any imbalance or change in these climate regimes over short or longer time spans can have a significant effect on the tropical rainforest, as well as climate conditions around the world.

Climate Change

Natural changes in climate have affected the Tropical Rainforest Biome throughout time. Recent studies have challenged earlier thoughts that the rainforest has been only weakly affected by climate changes since its origin in the Tertiary (65 mya). New studies document that small decreases in mean temperature between 1.8 million and 12,000 years ago caused significant changes in forest stands.

When cooler temperatures prevailed, water vapor in the atmosphere decreased, and the climate became drier than today (see the sidebar on p. 24). Alternatively,

Past Climate Change in the Tropics

Tropical rainforests have expanded and contracted over the last 10 million years due to changes in climate. Pollen records show several episodes of montane and savanna expansion, while rainforest disappeared from marginal areas. Fossil evidence indicates much drier and cooler conditions than present occurred over most tropical and equatorial areas. During this time, the rainforests of Africa and the Neotropics continued to contract into smaller fragments now called refugia. During interglacial cycles rainforests expanded, and species that survived in these refugia were able to radiate to fill newly available niches. These cycles of cooling and warming continue to affect the distribution of rainforest worldwide.

Geomorphic research on rising and falling lake levels in tropical areas confirm what fossils and pollen suggest. Lake levels decreased significantly during the glacial maximums. They rose during the interglacials. Fallen lake levels and drier climate contracted the rainforests of Africa, the Neotropics, and Australia, but had different effects on Southeast Asia. Lower sea levels during these times exposed extensive land bridges throughout the islands of the China Sea. This allowed for the widespread dispersal of plants and animals throughout the region.

The last glacial cycle occurred 18,000–12,000 years ago, and a significant decrease in rainforest species is evident during that time. After this last glacial cycle, warmer and wetter conditions characterized the early Holocene (10,000–5,000 years ago) and caused the rainforest to expand slightly. By about 5,000 years ago, humans had begun to alter the landscape significantly by clearing forests for firewood, charcoal, and agriculture. These activities may have affected rainforest distribution even more than climate change.

warmer temperatures correlate with greater water vapor content, and consequently higher rainfall occurs with warmer climates. Changes in vegetation relate to these climatic shifts. Current and future climate change, caused by high levels of greenhouse gases emitted into the atmosphere, will continue to cause changes within the Tropical Rainforest Biome. Climate change scientists predict various scenarios for tropical rainforests, from decreased precipitation, prolonged drought, increasing temperatures, and seasonality leading to severe desertification in the Amazon and African rainforests to increased rainfall—by as much as 50 percent—in the tropical forests of Asia. Any of these changes could affect the fragile balance of the Tropical Rainforest Biome and lead to its destruction. Add to these potential changes the direct changes caused by deforestation, fire, agriculture, mining, population pressure, and other exploitive events, and the future of the tropical rainforest may be filled with profound ecological changes. As the Tropical Rainforest Biome has great influence on the world's climate, its potential demise could bring with it catastrophic effects throughout the globe. Further research, conservation, and a decrease in the factors causing global climate change are necessary to ensure the future survival of this biome and the stability of the biosphere as a whole.

Soils

Soil is a major component in the creation and maintenance of vegetation in tropical biomes; and similarly, vegetation, along with climate, parent material, topography, and time, influence soil types and characteristics. Soil is formed by the weathering of rock from parent material (the geologic formation in the region) and the addition of dead and decaying organic matter from plants and animals. Weathering is the wearing away of this parent rock into smaller and smaller particles by chemical or mechanical processes. Weathering and soil formation are facilitated by climate (temperature, rainfall, and humidity) and the colonization of microbes and simple plants. These initial soil-forming processes are combined with the continuous addition and buildup of decomposing organic matter, mixing with small rocks and soil fragments. Soil-forming processes are lengthy and complicated.

Tropical soils are complex and varied. They typically are highly acidic, older soils with low fertility. Millions of years of constant heat and rainfall have shaped these soils. Older, more acidic, and less fertile soils are dominant in South America and Africa. The more fertile, younger soils share dominance in tropical Asia.

Soil Types

The most widespread tropical soils are ancient and deeply weathered oxisols and ultisols. Most of the remaining soils are more recently derived from volcanic activity and the breakdown of volcanic rocks, where weathering has only begun.

Tropical soils are primarily oxisols, as identified in the U.S. Soil Taxonomy Classification. The United Nations Food and Agriculture Organization (FAO) soil classification uses the term ferralsols. Latosol or laterites are other names applied to these tropical soils. Many of the tropical regions of the world have large areas of oxisols, particularly South America and Africa. Millions of years of weathering, facilitated by high rainfall and constant heat, created these ancient soils. Oxisols have little to no organic or humus soil layer. The constant heat, abundant moisture, and rapid bacterial decay prevent the accumulation of humus. Because of intense chemical weathering and abundant rainfall, all soluble minerals, and even some that are relatively insoluble, are leached. As a result, the A and B horizons may be tens of feet thick and indistinguishable from one another. Iron and aluminum are often high in oxisols. The excessive rainfall leaches out silica and important minerals and nutrients. In some regions, the high acidity of the soil combined with rainfall can lead to the release of soluble aluminum from the soil, making it toxic. Oxisols are very porous and have very low fertility. Their potential for erosion is low under natural conditions. Exposed to the tropical sun without the cover of decaying organic matter or plants, oxisols can become dry, very hard, and brick like. This process has been called laterization. This brick-like soil continues to be used as a building material for houses in the tropics.

Oxisols in the tropics have a characteristic deep yellow or red color due to high iron, or can be lighter due to higher aluminum content (see Plate I). They can also contain quartz, kaolin, and small amounts of other clay minerals. Bauxite (aluminum ore) is found within these soils in some areas and is mined, often with disregard to the devastation left in the aftermath. Oxisols are typically found in low-lying or flat areas within the tropics, but they occasionally can be seen on slopes. Oxisols are most prominent in South America, where two-thirds of all tropical oxisols are found. They appear mainly in the central and eastern Amazon Basin, derived from ancient Gondwanan basement rock. They also occur along the west coast of Colombia. In Africa, oxisols are concentrated in Burundi, Rwanda, the Democratic Republic of Congo, Cameroon, Gabon, and eastern Madagascar. They too have developed on the ancient Gondwanan basement rock. A small percentage of oxisols occur in patches in Sumatra, Java, Malaysia, the Philippines, and Thailand. With care, cultivation of rubber trees, coconut, tea, and cacao has been successful on these tropical soils.

Another type of tropical soil is ultisols, or acrisols and dystric nitosols (using FAO classification). They were formerly called red yellow podzolics. Ultisols are similar to oxisols chemically. They are highly acidic to acidic soils with high aluminum and iron content and have undergone extensive and long-term weathering. Where these soils are highly acidic, aluminum is converted to a soluble state, making the soils toxic. Ultisols are low in nutrients, with few minerals, and little to no humus or organic layer. Like oxisols, ultisols have high leaching potential. Unlike oxisols, ultisols often develop a thick, hard clay layer at depth, which increases their susceptibility to erosion. Ultisols are red and yellow, well drained, and typically found on slopes. The parent rock is not part of the ancient geologic shield that created the oxisols of Africa and South America. Ultisols are most dominant in tropical Asia, occurring in China, Thailand, Laos, Vietnam, and Indonesia. Ultisols also occur in the Amazon Basin west of Manaus, and along the eastern coast of Central America and Brazil. They are present in parts of tropical western Africa in Sierra Leone and Liberia. Ultisols are minimally suitable for agricultural conversion. Shifting or slash burn agricultural practices with long time intervals of fallow can retain the soil and its limited fertility. Short fallow intervals, increasingly common with rising populations and increased demand for food, turn these soils into infertile landscapes needing decades or centuries to recover.

On younger substrates, particularly those of volcanic origin, tropical soils may be quite fertile. Soils within the taxonomic order inceptisols ("beginning soils"), also called andepts or andosols (according to the FAO classification), are the third most widespread soil type in the tropics. These soils originated as volcanic ash. Two types of ash are present. Andesitic ash is high in nutrients; rhyolitic ash is less fertile and high in silica. Soils from andesitic ash are rich in organic matter and have a high water-holding capacity. Because they are rich in organic matter, they tend to be black or brown in color. Chemically, they have a high capacity to fix phosphorus, making it unavailable for plant use, and this can constrain vegetative growth. When they occur on slopes, they are susceptible to erosion by water and wind. Inceptisols play an important part in the vegetation of the humid tropics of the Philippines, Papua New Guinea, and Indonesia. This soil also occurs in parts of Central America, the Caribbean, and Ecuador, as well as in the highlands of Central Africa where volcanism has dominated earlier landscapes. As these soils are quite fertile and easy to till, many of the areas with these soils have been converted to agriculture, growing coffee, tea, and cacao in most regions; coca in South America; and rice in the Asian Pacific (see Plate II).

Many other soil types occur throughout the Tropical Rainforest Biome. Entisols (fluvents or lithosols) are soils in early development and are found on recent alluvial plains, shallow soils on steep slopes, and to a lesser degree, on fertile deep sands. Wet inceptisols (also called aquepts or gleysols by FAO) occur on older alluvial plains and along rivers and swamps in tropical Central and South America, Africa, and Asia. These permanently wet soils have been used for rice cultivation for centuries in Asia.

Other soils in the tropics include the more fertile alfisols, fluvents, lithic entisols, and spodosols. There are a few other soil types found within the Tropical Rainforest Biome, but they occur over small areas.

Soil Characteristics

Soil temperature shows even smaller variation than air temperatures in the tropical rainforest. Ground temperature is influenced by soil moisture content as well as air temperature. Within the root zone, about 2–20 in (5–50 cm), temperatures hardly fluctuate. There appears to be no diurnal or annual cycles for ground temperatures in the rainforest. These constant temperatures allow tropical plants to remain physiologically active throughout the year. This is different from mid-latitude soils and biomes where soil temperature fluctuations have a major influence on plants' metabolic activity.

Soil moisture can be a determining factor in the formation and stability of the tropical rainforests. Higher water storage capacity of the soil is crucial in tropical rainforests, particularly where rainfall is less constant and can lead to a deficit in the water balance of the ecosystem. Soil water storage capacity is closely related to soil texture. Tropical soils tend to be largely clay. Clay soils tend to be low in nutrients and fertility, and have a high water-holding capability. Superstaturation of some of these clay soils on steeper slopes can lead to erosion and landslides. A smaller, yet significant percentage of tropical soils are considered loams. Loamy soils have higher nutrient and organic content than clays. These soils are more fertile, and they allow for deeper root penetration and consequently fewer tree fall hazards. Few tropical soils are sandy with the exception of smaller areas within the Amazon and Congo basins and in the mangrove forests of the world.

Topography is an important component in soil formation and vegetative growth in the Tropical Rainforest Biome. Flat to gently sloping hills (0–10 percent slope angle) occupy about a third of the tropical landscape. These areas are typically well drained. A small percentage of these flat areas have soils with poor drainage and are swamps or flooded forests. An additional third of the tropics are rolling hills (10–30 percent slope). The last third are steep slopes (>30 percent slope). The steeper slopes tend to have shallow soils and are often rocky.

Nutrient Cycling and Decomposition

Tropical soils contribute little to the cycling of nutrients within the Tropical Rainforest Biome; however, some contribute needed phosphorus and nitrogen. The larger contributor is the forest vegetation and organisms themselves within the layer directly above the soil. Intense organic activity decays material dropped from plants and dead organisms. This decomposing layer can be 6–8 in (15–20 cm) in depth. It is the lifeblood of the Tropical Rainforest Biome. The rapid decomposition of dead plants and animals is undertaken by many organisms, from insects to smaller bacteria and fungi. Scientists have just begun to research the bacteria and fungi that are so essential in the maintenance of the rainforest.

Rainforest vegetation has adapted to soils with little fertility by rapidly recycling nutrients from decomposing material in the forest litter. Beneficial aerobic and anaerobic bacteria convert unusable compounds into usable minerals and nutrients that are necessary for plant growth. The uptake of decomposed, mineralized matter is facilitated by roots and their accompanying beneficial fungi (mychorrhiza) that help the plants utilize nutrients. The roots of tropical plants tend to be shallow and often above the ground to allow for the greatest use of these nutrients. Small complex roots form networks with mychorrizal fungi that rapidly absorb the nutrients and makes them available to the plant. In exchange for these services, the plants provide the microorganisms with food and shelter among the roots. These microbes also have been found to help a tree resist drought and disease.

Vegetation

This ancient biome occurs under optimal growing conditions: abundant precipitation and year-round warmth. With no annual rhythm to the forest, each species has evolved its own flowering and fruiting season.

Forests have evolved to accommodate many species of various shapes, sizes, and heights that fill in the forest. Tropical broadleaf evergreen trees are the most common growthform (see Plate III). Sunlight is a limiting factor in the lower canopy and a variety of strategies and growthforms have been successful in the struggle to adapt to varied light, above and beneath the canopy. A vertical stratification of three canopy layers of trees is usual. Along with these tree layers is an accompanying array of woody vines, orchids, bromeliads, and epiphytes. The tree layers have been identified as A, B, and C layers with a shrub layer and ground layer below (see Figure 2.6).

The emergents are in the A layer. These are widely spaced trees 100–120 ft (30–36 m) tall with umbrella-shaped crowns that extend above the general canopy of the forest. These trees contend with strong drying winds and tend to have small leaves. Some may be deciduous during the brief dry season. The next layer, the B layer, is composed of a closed canopy of 60–80 ft (18–24 m) trees. Light is readily available at the top of this layer but greatly reduced below it. The C layer is often the third and last layer of the canopy. The C layer is composed of smaller trees around 30–60 ft (9–18 m) tall. This layer along with the B layer creates a closed canopy with little air movement and high humidity.

Below these three layers is the shrub and sapling layer where little light penetrates through the canopy layers. Less than 3 percent of the light reaching the top

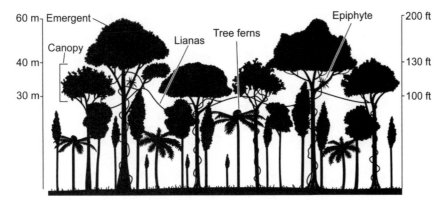

Figure 2.6 Forest structure of a lowland rainforest in the Neotropics and Asian-Pacific. The forest structure of African rainforest is similar but trees tend to be smaller. *(Illustration by Jeff Dixon. Adapted from Lambertini 2000.)*

of the forest canopy passes to this layer. Suppressed growth is characteristic of the young trees that inhabit this zone, although they are capable of a rapid surge of growth when a gap in the canopy opens above them. Shrubs that can grow in low light are found in this layer, many having large leaves to catch whatever light is available.

Little grows on the forest floor. Less than 1 percent of the light that strikes the top of the forest penetrates to the forest floor. In this shade, few green plants can grow. Moisture is also reduced by the canopy above: one-third of the precipitation is intercepted before it reaches the ground. The layer is composed of a few ferns and herbaceous plants, but mostly dead, decaying plants and animals and the organisms that decompose them (see Figure 2.7).

As the layers of vegetation in the canopy are varied, so too are the habitats they provide for other organisms. Each layer has a unique set of lifeforms and lifestyles of organisms that have adapted to the specific resources available to them.

Common Characteristics of Tropical Trees

Trees in the Tropical Rainforest Biome are often different from trees found at other latitudes, yet are similar throughout the tropical forest regions of the world. Some of the common characteristics include smooth bark or bark with spines or spikes, buttressed trunks, large leaves, and leaves with drip tips. In addition to similar tree characteristics, the tropical rainforests of the world contain climbing woody plants in the form of vines or lianas.

Buttresses. Many emergent trees found in the Tropical Rainforest have broad, woody flanges at the base of the trunk. These buttresses were originally believed to help support trees with a shallow network of roots, adding support in wet soils. Recently it has been shown that these buttresses also take part in carbon dioxide

Figure 2.7 Little vegetation grows on the forest floor of Madidi National Park, Bolivia. *(Photo by author.)*

and oxygen exchange, as well as channeling stem flow and dissolved nutrients to the roots. Buttresses can increase the surface area of a tree so that it can "breathe in" more carbon dioxide and "breathe out" more oxygen. Buttresses can be 15–32 ft (5–10 m) high where they join into the trunk. (see Figure 2.8).

Large Leaves and Drip Tips. Large leaves are common among trees of the C and shrub layers. Young trees destined for the A and B layers may have large leaves. The large leaf surface helps intercept light in the sun-dappled lower strata of the forest. If these trees reach the higher canopy layers, their newer leaves will be smaller. Leaves of tropical rainforest plants often come to a point at the top or end of the leaf. These are called drip tips, and they ease the drainage of rainfall off the leaf and promote transpiration (see Figure 2.9). They occur in the lower layers and among the saplings of species of the emergent layer (A layer).

Thin Bark. Another characteristic that distinguishes tropical species of trees from those of temperate forests is exceptionally thin bark, often only 0.02–0.07 in (1–2 mm) thick. In temperate forests, trees need thick bark to limit water loss through evaporation. Since moisture is not a limiting factor within the rainforest, thick bark is not necessary. The bark is usually smooth, making it difficult for other plants to grow on its surface. A few tropical trees are armed with spines or thorns to detract herbivores (see Figure 2.10).

Figure 2.8 Buttresses are common on emergent trees in the tropical rainforest. *(Photo by author.)*

Figure 2.9 Drip tips at the tip of the leaf allow the plant to quickly drain the abundant rainfall. *(Photo by author.)*

Figure 2.10 Many rainforest trees develop spikes or thorns on their trunk to discourage herbivores. *(Photo by author.)*

Cauliflory. A characteristic of many rainforest trees and vines not commonly found in other biomes is cauliflory. Cauliflory is the production of flowers on leafless trunks, rather than on twigs or smaller branches (see Figure 2.11). In many tropical trees and lianas, the flowers (and later fruits) are attached to short leafless stems or spring directly from the trunk or large branches. Most of the smaller branches and often the entire crown remain vegetative. There are several types of cauliflory. Some plants only produce flowers on their main trunk. This is called trunciflory. In others, flowers are limited to the base of the plant, called basiflory.

Most of the plants that display cauliflory are small to medium height, usually 16–65 ft (5–20 m), and belong to the lower canopy or shrub layers. The flowers of cauliflorous trees are often pollinated by birds and bats. Fruits are consumed by larger animals that may be unable to reach fruit in the canopy. These trees need these animals for the dispersal and germination of their seeds. Some trees exhibiting this adaptation develop fruits that are too heavy to be supported by branches. Many trees in the fig family (Moraceae) display cauliflory.

Survival below the canopy depends on the plant's ability to tolerate constant shade or adapt strategies to reach sunlight. Shrubs, vines, epiphytes, and nonphotosynthetic plants have developed specific strategies to reach or receive sunlight or to do without it. Vines are an important structural feature within the tropical rainforest accounting for a large portion of the forest biomass and competing for water, light, and nutrients. Their fruits are a primary food source for tropical animals.

Figure 2.11 The fruit growing from the trunk of this cacao tree is a type of cauliflory. *(Photo by author.)*

Vines can be danglers, stranglers, or climbers. Lianas are dangling woody vines that grow rapidly up the tree trunks when a temporary gap opens in the canopy. Lianas typically start out as a shrub in the forest understory. As the tendrils from the branches intertwine around neighboring trees, the liana travels upward into the tree crown (see Figure 2.12). They flower and fruit in the tree tops of the higher canopy layers; many are deciduous. Lianas can spread through several trees while their stems remain rooted in the ground. As well as competing for light, water, and nutrients, lianas can often burden a tree with extra weight, making it vulnerable to toppling in high winds. Lianas are often found dangling from the branches of canopy trees.

Other vines are termed bole climbers. These vines also germinate on the forest floor and send out tendrils that grow around the trunk of trees. They develop aerial roots, using the tree as their anchor. These climbers may fully engulf the tree. Often, many species of bole climbers can inhabit one tree. Many climbers, including

Figure 2.12 The rainforest is abundant with woody vines throughout the canopy. Taken on São Tomé Island, Republic of São Tomé and Principe. *(Photo courtesy of Robert Drewes, Ph.D., California Academy of Sciences.)*

the ancestors of the domesticated yams (Africa) and sweet potatoes (South America), store nutrients in roots and tubers.

Competition for sunlight by plants can be deadly. Some vines are considered stranglers as they quite literally strangle the host tree. Strangler vines need sunlight to grow and reproduce. Seeds falling to the ground quickly die in the deep shade and infertile soil of the tropical rainforest. Stranglers begin life as epiphytes with their seeds deposited on the branches of a host tree by birds and small animals that have eaten their fruit. The seeds sprout and send long roots to the ground. The roots rapidly increase in diameter and successfully compete for water and nutrients in the soil. As the strangler matures, branches and leaves grow upward, creating a canopy that blocks sunlight. In time the tendrils, now vine-like, fuse with each other effectively surrounding and killing the host tree. Additionally, roots are sent out and wrap around the roots of the host tree competing for nutrients (see the sidebar on p. 35). Many stranglers are fig trees (Moraceae family); other genera of stranglers include pitch apple (Clusiaceae), umbrella tree (Araliaceae), tree jasmine (Rubiaceae), and Coussapos (Moraceae). Some do not kill their host tree.

Epiphytes. Epiphytes (also called air plants) live on other plants. Typically, they are not parasitic, although they do compete for resources. Epiphytes attach to tree branches and use the soil and dust particles they trap within the canopy to

Competing for Light at Any Cost

Strangler figs of the Moraceae family are abundant in all tropical rainforest regions. They provide an important food source for many tropical animals. Each fig species has a specific wasp that is its exclusive pollinator. Strangler figs reach high into the canopy and can become trees of the emergent layer, competing with the host tree for space and light. If the strangler wins, the host tree will eventually die because of shading, constriction of bark, or competition for resources. Once dead, the host tree will rot away, leaving what looks like a tree with a hollow inside, and the strangler, once an epiphytic fig plant now becomes an independent, often majestic, tree (see Figure 2.13).

Figure 2.13 Strangler figs will often outcompete their host tree for resources. When the host tree dies, a hollow space is left. *(Photo by author.)*

grow. These particles are the source of calcium, phosphorus, potassium, and other minerals and nutrients needed for plant growth. Epiphytes can develop a root mass within the canopy that serves as a reservoir for moisture and nutrients often used by the host tree as well. Mychorrizal fungi are also present in the root mass and assist the epiphyte in nutrient absorption. An amazing variety of lichens, mosses, liverworts, ferns, cacti, orchids, and bromeliads, to name a few, are epiphytic. Species diversity studies have found that a typical tropical canopy tree has at least 50 kinds of epiphytes living on it. Some epiphytes even grow on other epiphytes.

Nonphotosynthetic plants. Nonphotosynthetic plants are heterotrophs—organisms that need food from other organisms—that live on the forest floor. Some are parasites that derive their nutrients by tapping into the roots or stems of photosynthetic species. *Rafflesia arnoldi*, a root parasite of a liana, has the world's largest flower, more than 3 ft (91 cm) in diameter. It produces an odor similar to rotting flesh to attract pollinating insects. Saprophytes, more correctly called saprotrophs, can be nonphotosynthetic plants, fungi, and bacteria that derive their nutrients from decaying organic matter. Other nonphotosynthetic plants are parasitic, such as some orchids that use mychorrhizal fungus to facilitate food absorption.

Plant adaptations have already been mentioned in terms of life strategies and placement in the canopy layers. Tropical trees and shrubs have developed adaptations to expedite germination or seed dispersal as well. Some have large, fleshy fruits that attract birds, mammals, and even fish as dispersal agents. The animals will digest the flesh, leaving the seed to germinate in a new location. Plants growing in flooded forests have fruits that float. Additional plant adaptations include strategies that enhance reproduction, or in some cases, predation. Many tropical plants have evolved a mutualistic relationship with their pollinator either in terms of attracting them to the flower through color, scent, or nectar, or by mimicry—looking like the pollinator's mate, for example. Carnivorous plants create sweet or bad smelling nectar that is attractive to their prey.

Root Systems

The forest may be the most obvious part of the biome, but what underlies the immense and lush trees is a system of roots that support the trees, the backbone of the forest. Little work has been done on the root systems within the tropical rainforest; however, some basic types of roots and their roles within the rainforest are understood. Aerial roots are common among tropical plants. They can provide a measure of stability in unstable substrates and are a means of aboveground aeration in water-soaked soils. For stranglers, aerial roots provide initial anchorage and subsequent feeding and support.

Symbiotic or mutualistic associations of the root systems of trees with soil microorganisms have been receiving more attention by researchers. The process of recycling nutrients from the decaying material into the plant is largely dependent on the activity of soil microbes, which aid in decomposition and in root absorption of the nutrients. Complex shallow roots form a wide network with microbes providing a larger surface area to absorb the resources necessary to maintain the tallest of trees.

Plants throughout the tropics have developed similar adaptive strategies. The structure and characteristics discussed above are evident in each of the regional expressions of the Tropical Rainforest Biome.

Animals

Studies of tropical rainforest animals reveal an overwhelming diversity of species compared with other terrestrial biomes. As many as 450 bird, 93 reptile, 37 amphibian, and 70 mammal species have been recorded on less than 247 ac (1 km^2) in the Guyanan rainforest in South America. Other tropical regions of the world show similar diversity. Isolated tropical rainforests and those of more recent origin, such as those of the Hawaiian and Pacific Islands, Australia, and New Guinea, reveal less variety. Studies on butterflies have reported species counts as high as 300–500 in a given study area of tropical rainforests of Africa and South America. Termites and beetles show even higher numbers.

The tropical rainforest landscape is varied in structure, growthforms, and species. This diversity of habitats, along with diverse survival strategies in the use of those habitats, provides opportunities for many more animals than the homogeneous habitats of more simply structured biomes, such as grasslands. Partitioning of space, food resources, and timing of activity expand the species capacity of the forest. Diurnal and nocturnal behaviors allow different animals to use the same habitat at different times. Using different layers of the canopy as habitat further separates resources and increases capacity. Selection of varied food choices, such as leaves, gums, bark, fruits, nuts, seeds, and other organisms, within the same environment, allows for the survival of many species without direct competition. Even the location of food sources can be partitioned to provide diversification. For example, on the island of Trinidad in the Caribbean Sea, three species of tanagers have been found to extract insects from different microhabitats off the same tree without directly competing. One species of bird feeds on insects from the underside of leaves, another from small vines, twigs, and leaf petioles, and the third takes insects from the main branches. This allows all three species to utilize the same food source without the need to compete.

Common characteristics found among mammals, birds, reptiles, and amphibians include various adaptations to an arboreal life. For example, arboreal anteaters, pangolins, some rodents, and many monkeys and possums in the rainforest have developed prehensile tails. A prehensile tail allows the animal to hang on to a tree branch while using its other limbs for catching prey, eating, or grooming. Another adaptation to arboreal life is the evolution of opposable digits that aid in climbing. Many arboreal species also have sharp claws to hang on to trees and branches. Additional adaptations include binocular vision to aid in accurately seeing distances and dimensions. This is particularly helpful when leaping between trees within the dimly lit forest.

Bright coloration in birds allows them to blend in with the brightly colored flowers and fruit that have evolved coloration to attract pollinators. It also helps with species recognition when birds seek mates in the dense foliage of a vast forest. Various beak sizes allow birds to utilize different food sources, such as fruits, nuts, insects, and even crabs.

Yet another adaptation to arboreal life is the development of limbs or skin flaps that aid in gliding through the forest. This is particularly varied in Borneo. Here gliding squirrels, colugos, snakes, lizards, and frogs travel throughout the forest canopy seldom visiting the forest floor. Loose skin along the body and limbs spreads out to form a broad surface, aiding in long-distance gliding through the forest.

Loud vocalization in birds and primates is an adaptation for communicating within the dense forest layers. The howler monkey of the Neotropical Rainforest and siamang in Asia are known for their long-distance calls and the variety of vocalization within and outside a group. These primates are most vocal at dusk and dawn. The shrieks of Scarlet Macaws can be heard far below and away from the flock.

Some tropical animals exhibit warning coloration; this is also called aposematic coloration. Warning coloration makes a dangerous, poisonous, or foul-tasting animal particularly conspicuous and recognizable to a predator. Several animals exhibit this coloration, such as the yellow and black stripes of bees and wasps, orange and black coloration in beetles, and the bright red or yellow colors of many poisonous frogs and snakes. Warning coloration alerts potential predators that the species is dangerous or toxic and should not be eaten. Would-be predators learn to avoid organisms with warning coloration.

The survival of any tropical animal depends on its ability to find food, shelter, water, and mates, and to avoid predation. One of the adaptive strategies found in tropical animals is mimicry. Mimicry is a form of biological interaction in which species resemble the appearance or behavior of other species. Several types of mimicry are common. Similarity in appearance, or so-called parallel mimicry, signals or warns potential predators that an animal is unpalatable or poisonous by having several different species adopting the same warning colors. The use of similar coloration, particularly reds, oranges, and yellows, is found among beetles, butterflies, and other insects, and serves to warn potential predators. Batesian mimicry is the condition when a nonharmful species mimics the appearance of a harmful species within the same family to avoid predation. This strategy is common among tropical animals. The Monarch butterfly is mimicked by the nonpoisonous Viceroy butterfly, a classic example of Batesian mimicry. This type of mimicry is also used by some less harmful snakes so that they look highly poisonous and predators will leave them alone. Colors and patterns of the harmless false coral snake and the mildly poisonous coral snake are similar to those of the very poisonous coral snake. Other organisms display predatory mimicry and resemble their prey, increasing their chances of a more successful marauding venture. Plants also possess mimicry strategies, their flowers often mimicking the scent or appearance of a potential pollinator's mate or, in the case of carnivorous plants, their prey.

An adaptation similar to mimicry is camouflage, the development of an appearance that blends in with the environment to escape notice or predation, or to allow for effective hunting by a predator. Predators can move close to their prey if they can remain unseen. The opposite is also true; camouflaged prey animals can be difficult to detect by a predator. Animals that live in rainforests may be green or brown to match the colors of foliage and tree trunks. Many insects use camouflage to avoid predators. Stick insects mimic branches or twigs and move slowly or remain still for long periods. Some may develop wings that look like leaves, complete with veins (see Plate IV). Grasshoppers, other leaf hoppers, and spiders can be green or brown to blend in with their environment. Frogs also use camouflage, developing the coloration of their primary environment. Chameleons of tropical Africa and Madagascar are the experts at camouflage. Chameleons are reptiles with the ability to change their color rapidly as they change habitats or moods. The South American anole lizards, less spectacular color-changers, exhibit similar behavior.

Animals of the Tropical Rainforest Biome are incredibly diverse. These species have evolved to filled myriad niches available in the varied canopy layers. They use the abundant resources available to them. By uniquely adapting, animals and plants have continued to thrive in the Tropical Rainforest Biome. In the different regional expressions of the Tropical Rainforest Biome, these adaptations are evident, some taken on by entirely different kinds of species. The diversity within the forest allows great numbers of species to exist. Though species numbers are high, the actual population numbers are low. Extinction is high in the tropics when species populations become restricted to small areas, due to deforestation and fragmentation of the forest. Such destructive processes in even small areas can wipe out entire species.

Tropical Vertebrates

Mammals are abundant and diverse in the Tropical Rainforest Biome. Primates including marmosets, monkeys, and great apes are found there and a number of them spend their entire lives in the trees. Large rodents, such as the pacas, and capybaras of the Neotropical Rainforest, provide food for animals and humans. Smaller rodents such as mice, rats, agoutis, squirrels, and porcupines among others take advantage of the varied vegetation and canopy layers. Insectivores, such as anteaters, armadillos, and sloths, are confined to the Neotropics, while pangolins, lemurs, and marsupials fill these niches in the forests of Africa and the Asian Pacific. Carnivores such as jaguars, panthers, leopards, and tigers, as well as weasels, mongooses, and civets are the main predators of the forest.

Bats are a critical component of tropical ecosystems. Tropical rainforests contain more kinds of bats than any other biome. They serve the rainforest through plant pollination, seed dispersal, and insect predation. Bats are divided into two suborders, Megachiroptera and Microchiroptera. Megachiroptera are primarily large bats that feed on fruits and nectar and are often called flying foxes. They are restricted to the Old World, that is, the regions of Africa, Asia, and Australia.

Microchiroptera are distributed worldwide. They are primarily insect eaters, but also feed on fish, fruit, nectar, and blood.

The tropical rainforests of Australia and New Guinea, and to a lesser extent, South America, house an additional group of mammals, marsupials. The Austral-asia marsupials evolved and diversified after this region separated from the rest of Gondwana. This isolation allowed for vast adaptive diversification found nowhere else in the world. Marsupials found in the Neotropics are probably of more recent origin.

Tropical rainforests have the highest diversity of birds of any biome; however, the families represented and actual species found vary considerably among regions. The Neotropical Rainforest is by far the richest in bird species, followed by the Asian-Pacific and African regions. Different taxa (evolutionary groups) have colonized and diversified in different geographic areas. Greater evolutionary age, diversity of habitats, abundant resource, and the unique diversification of feeding specializations have supported this great diversity of species. Because of the numer-ous canopy layers, year-round availability of flowers, fruits, insects, nectar, and other food sources, birds in the tropical rainforest show a vast array of forms and behaviors. Ground-dwelling birds take advantage of the low-growing vegetation, fallen fruits, insects, and other organisms in the fallen leaves and debris. These birds include guans, curassows, tinanous, and chachalacas from the Neotropics; peafowl, guinea fowl in Africa; and doves, peacocks, partridges, bowerbirds, and cassowaries in the Asian Pacific.

Brightly colored birds such as parrots, macaws, toucans, and hummingbirds of South America; touracos, starlings, sunbirds of Africa, cockatoos, and the amazing birds of paradise found on Australia and New Guinea are some of the birds that are flashes of brilliant colors as they fly through the forests. Other birds, such as weaver birds of Africa and oropendolas of Central and South America, make their nests as hanging baskets amid the tropical tree branches. A large number of other birds, including predators such as eagles, goshawks, hawks, falcons, and owls com-plete the diverse assemblage of tropical birds.

Reptiles and amphibians of the tropical rainforests, including snakes, lizards, crocodiles, and frogs and toads, among others are found on the forest floor, as well as high in the canopy. Tropical snakes include venomous snakes such as pit vipers, cobras, and coral snakes, and constricting snakes such as pythons and boas (including the anaconda), as well as reticulated and rock pythons, the largest snakes in the world. Vine snakes and other tree snakes also travel through forest lit-ter, tree branches, and among leaves, as well as in waterways. These are important components of life and death in the rainforest.

Many lizards, including skinks, iguanas, geckos, basilisk lizards, and monitor lizards (the largest lizards in the world), are found in tropical rainforests. Chame-leons are found in Africa and Madagascar. Turtles and crocodiles inhabit the riv-ers, swamps, and flooded forests.

Frogs are the most abundant amphibians in the rainforest. Tropical frogs are most abundant in the trees, and relatively few are found near bodies of water or the forest floor. Frogs must always keep their skin moist since most of their respiration is carried out through the skin. The high humidity of the rainforest and frequent rainstorms give tropical frogs infinitely more freedom to move into the trees and escape the many predators of rainforest waters. The majority of rainforest frogs place eggs in vegetation or lay them on the ground. By leaving the water, frogs avoid egg-predators like fish, shrimp, aquatic insects, and insect larvae. Frogs have evolved to take advantage of the many layers of the tropical rainforest, some live and breed high in the canopy in bromeliads, others have developed toe pads to climb and hang onto vegetation. Many toads and frogs, such as the poison dart frog of the Amazon, have developed toxicity to discourage predation.

Humans play a role in tropical species presence and distribution. Regions where humans have been present for millions of years seem to have fared better in terms of the presence of larger animals. Populations of terrestrial vertebrates in those areas more recently inhabited, such as the Neotropics, Madagascar, and New Guinea, have decreased or gone extinct. In their travels and migrations, humans have transported (both intentionally and unintentionally) species foreign to the new area. These newcomers often adapt and do well in their new environment to the detriment of native species of the region.

Tropical Invertebrates

By far the most abundant and successful animals of the tropical rainforest are insects, as demonstrated by their tremendous diversity. In the rainforest canopy, insects abound. A study of the rainforest canopy in Peru found more than 50 species of ants, 1,000 beetles species, and 1,700 other species of arthropods, including more than 100,000 individuals in a plot measuring 30 in^3 (500 cm^3). A rainforest tree alone can have some 1,200 species of beetles, while a 2.4 ac (1 ha) plot of rainforest canopy is projected to have more than 10 times that number.

Additionally, many insects and other organisms not typically found in the trees make their home in the canopy of the tropical rainforest. Several species of crab have been found hundreds of feet above the ground in bromeliads of Neotropical rainforests. Similarly, earthworms and giant planarians (flatworms) are also part of the canopy system. Earthworms play an important role in the processing of canopy soils and mulch that supports epiphytic growth. Even leeches are found in the forest canopy. Mosquitoes are abundant in the canopy with fewer on the ground. Many insects like stick and leaf insects, katydids, leaf hoppers, and mantids have developed incredible behavior, body structure, or color to adapt to their surroundings.

Products and Services from the Rainforest

Hundreds of items used in the world today come from tropical rainforests. They include fruits, vegetables, spices, cocoa, coffee and tea, oils, cosmetics and perfumes, houseplants, fibers, building material, and medicines (see Table 2.1).

Table 2.1 Products from the Tropical Rainforest

Tropical Woods
Teak, Mahogany, Rosewood,
Balsa, and Sandalwood

Houseplants
Anthurium
Dieffenbachia
Dracaena
Fiddle leaf fig
Mother-in-law's tongue
Parlor ivy

Philodendron
Rubber tree plant
Schefflera

Silver vase bromeliad
Swiss cheese plant
Zebra plant

Spices
Allspice, Black Pepper,
Cardamom, Cayeene, Chilli,
Cinnamon, Cloves, Ginger,
Mace, Nutmeg, Vanilla,
Paprika, and Turmeric

Gums and Resins
Chicle latex (chewing gum)
Copaiba (perfume, fuel)
Copal (paints, varnishes)
Gutta percha (golf ball covers)
Rubber latex (rubber products)
Tung oil (wood polish)

Medicines
Annatto (red dye)
Curare (muscle relaxant for
  surgery)
Diosgenin (birth control, sex hor-
  mones, steroids, asthma, arthri-
  tis treatment)
Quassia (insecticide)

Fibers
Bamboo (furniture baskets, flooring)
Jute (rope)
Kapok (insulation, soundproofing, life jackets)
Raffia (rope cord, baskets)
Ramie (cotton-ramie fabric, fishing line)
Rattans (furniture, wickerwork, baskets, chairs)

Oils
Bay oil (perfume)
Camphor oil (perfume, soap, disinfectant,
  detergent)
Coconut oil (suntan lotion, candles, food)
Eucalyptus oil (perfume, cough drops)
Oil of star anise (scenting, infections, beverages,
  cough drops)
Palm oil (shampoo, detergent)
Patchouli oil (perfume)
Rosewood oil (perfume, cosmetics, flavorings)
Sandalwood oil (perfume)
Ylang-ylang (perfume)

Fruits
Avocado, Banana, Coconut
Grapefruit, Lemon, Lime, Orange
Mango, Papaya, Passion Fruit
Pine, Plantain, Tamarind, Tangerine

Vegetables and Other Food
Brazil nuts, Cashews, Macadamias, Peanuts
Cane sugar, Chocolate, Coffee
Cucumber, Hearts of palm, Okra, Peppers
Manioc/Tapioca, Mayonnaise
Soft Drinks, Tea, Cocoa, Vermouth

Quinine (antimalarial, pneumonia treatment)
Reserpine (sedative, tranquilizer)

Strophanthus (heart disease)

Strychnine (emetic, stimulant)

Source: Denslow and Padoch 1988.

Unsustainable harvesting of some of these tropical products and clearing the forest for plantation crops is a major cause of rainforest deforestation. Illegal logging and hunting causes tremendous loss of forest and species.

Some of the most important rainforest products are medicinal drugs. More than 25 percent of pharmaceutical products used in the United States are derived from tropical plants. Quinine is used to treat malaria and pneumonia; it was developed from the bark of the Neotropical chinchona tree. Curare, an ingredient in most muscle relaxants, comes from the Amazon. Other drugs, including contraceptives, those that stimulate the heart and respiratory systems, some that inhibit the growth of tumors, some that cure childhood leukemia, and some that are potential treatments for cancers and AIDS, have all been discovered in the tropical rainforest. More research is needed to identify the chemical properties of rainforest plants and other organisms and their effect on humans. Much knowledge is to be gained from the native people of the forest who have used these plants for medicinal purposes for thousands of years. Unfortunately, native cultures, along with the rainforest they live in, remain in jeopardy.

Human Impact

People have lived in the Tropical Rainforest a long time. They have made use of the abundant resources of the forest for food and shelter and have modified the forest for agricultural purposes for tens of thousands of years. While these populations were small, such uses were sustainable and had minimal permanent impact on the forests. However, as populations have increased in forest communities, and others have migrated to the forest and areas surrounding the forests, human impact has become significant. Burning and clearing of forests for agriculture and livestock production for both local and export markets have left some formerly forested areas barren, requiring hundreds of years to recover. Consumer demand for tropical woods and unsustainable timber harvesting practices have led to the virtual extinction of some tree species. Mining for oil, gold, copper, and other valuable resources has destroyed vast areas of forest. The processes used in extracting these resources have left the land and many tropical rivers toxic and unusable. Wars and diseases have depleted many indigenous people as well as animals. Hunting has also decreased animal populations. Although current estimates of rainforest destruction are varied, it is clear that in some areas rainforest destruction is rapid and ongoing. Of the Amazon's total coverage of 1.8 million mi^2 (3 million km^2) of rainforest, about 240,000 mi^2 (600,000 km^2) have been destroyed by human encroachment in the past few decades. Economic development of poorer nations, their increased debt to foreign investors, and continued and rapid population increases further threaten the future of the Tropical Rainforest Biome.

Efforts by local communities, governments, and international organizations are making some progress in slowing forest destruction. Dedicated people around the world are developing strategies to support economic growth without completely depleting the rainforests. Although significant destruction and disappearance of rainforests have occurred, continued efforts at conservation, and continued research into the ecology of the rainforests will be necessary to increase awareness of the importance of this incredibly diverse and important biome. Sustainable ways to use these tropical resources will be critical for tropical nations and their forests to survive.

Further Readings

Books

Bermingham, Eldredge, Christopher W. Dick, and Craig Moritz, eds. 2005. *Tropical Rainforests: Past, Present, and Future.* Chicago: University of Chicago Press.

Gay, K. 2001. *Rainforests of the World. A Reference Handbook.* Santa Barbara, CA: ABC-CLIO, Inc.

Lambertini, M. 1992. *A Naturalist's Guide to the Tropics.* Chicago: University of Chicago Press.

Newman, A. 2002. *Tropical Rainforest.* New York: Checkmark Books.

Primack, R., and R. Corlett. 2005. *Tropical Rain Forests.* Oxford: Blackwell Publishing.

Internet Sources

Adams, J. 1994. *The Distribution and Variety of Equatorial Rain Forests.* Oak Ridge National Laboratory. http://www.esd.ornl.gov/projects/qen/rainfo.html.

Butler, R. 2007. *Rainforests.* Mongabay.com. http://rainforests.mongabay.com.

3

Regional Expressions of the Tropical Rainforest Biome

Three major regional expressions of the Tropical Rainforest Biome are apparent: the Neotropical, the African, and the Asian-Pacific regions. Distinctions among them are based on location, origin, climate, soil, forest structure, and species composition. They exhibit similarities (and differences) related to geologic, climatic, evolutionary, and ecological events. Some plants and animals found in all regions of the Tropical Rainforest Biome reflect early origins prior to the breakup of Gondwana or a time when rainforests were widely distributed across the continents (see Chapter 2).

These and other environmental events of the past influenced the differences seen between the regions. Clearly, the isolation of the regions from one another provided time and opportunity for species to evolve unique adaptations to their respective environments, as is evident in the flora and fauna of New Guinea and Northern Australia. Orogenic events, such as mountain building, also isolated species and created avenues for further adaptive radiation. Episodes of severe climate change restricted the distribution of previously widespread species, isolating them and leading to the evolution of endemic species (restricted to the area of origin).

Each region has elements that make it unique within the Tropical Rainforest Biome. The Neotropical rainforest is the largest and most extensive, with the highest diversity of flora and fauna on the planet. Its long isolation produced endemic plants and animals uniquely adapted to the Neotropical environment.

The African rainforest is smaller in extent and species diversity than the other regions. The cooler and drier climate of the Pleistocene Epoch contracted the rainforest significantly. This constriction, along with the dominance of single species in

45

some areas, influenced the diversity of other plants and animals. Trees in the tropical rainforests of Africa tend to be shorter, and the forest is less dense than other regions. The African region shows the highest diversity of primates, as well as an abundance of large ground-dwelling mammals, such as elephants and ungulates.

The Asian-Pacific rainforests are distinctive due to the presence of the Dipterocarpaceae family that dominates the forest trees. These trees are among the tallest trees in the entire Tropical Rainforest Biome and occur in large clumps. Asian forests also display an abundance of primate species and gliding animals, while the rainforests of New Guinea and Australia contain an abundance of marsupial mammals and a diverse radiation of endemic, flamboyant birds.

Table 3.1 provides a few of the differences among the various regions in terms of elevation and major geological and biological features. Although each region has both a shared and a separate evolutionary history, many of the plants and animals resemble each other or fill similar niches. Adaptations to regional environments in distinct locations reveal similar habitats separated by thousands of miles and millions of years. Convergent evolution, mimicry, and adaptive radiation are all apparent in the three regional expressions of the Tropical Rainforest Biome.

The Neotropical Rainforest (Central and South America)

The Neotropical expression of the Tropical Rainforest Biome occurs within the tropics of Central and South America. Three main subregions of rainforest occur: the Atlantic and the Caribbean, the Choco in Colombia, and along the Amazon and its tributaries. Belize, Bolivia, Brazil, the Caribbean Islands, Colombia, Costa Rica, Cuba, Ecuador, El Salvador, French Guiana, Guatemala, Guyana, Honduras, Mexico, Nicaragua, Panama, Peru, Suriname, and Venzuela all contain parts of the Neotropical rainforest (see Figure 3.1).

In Central America, tropical rainforests occur in patches from the Yucatan Peninsula through Central America and into northern Venezuela. In this region, tropical rainforests are more abundant on the Caribbean side than on the Pacific. However, there are some large areas of rainforest on the Pacific side, particularly in Costa Rica. Remnants of tropical rainforest exist on the Caribbean islands of Cuba, Jamaica, and Hispaniola (Haiti and the Dominican Republic), and to some extent, on the Lesser Antilles and in Puerto Rico.

In South America, tropical rainforests are found along the Pacific coast from Panama and Colombia into northern Ecuador, in the area known as the Choco. The Choco experiences the greatest rainfall of all Neotropical subregions. The largest area of rainforest in the world lies on the eastern side of the northern Andes throughout the Amazon Basin. About the size of the 48 contiguous United States, the Amazon Basin covers 40 percent of the South American continent. The Basin is drained by the Amazon River, the world's largest river in terms of discharge, and the second longest river in the world. The Amazon River is made up of more than

Table 3.1 Comparison of Major Rainforest Regions

	Neotropics	Africa	Madagascar	Asia	New Guinea and Australia
Geographic area	Central America, Amazon River Basin	West Africa, Congo River Basin	Eastern Coastal forest	Malay Peninsula and islands	Large continental island
Annual rainfall in (mm)	80–120 in (2,000–3,000 mm)	60–100 in (1,500–2,500 mm)	80–120 in (2,000–3,000 mm)	>80–120 in (2,000–3,000 mm)	>80–120 in (2,000–3,000 mm)
Largest country	Brazil	Democratic Republic of Congo	Malagasy Republic	Indonesia	Papua New Guinea
Distinctive plant characteristics	Bromeliads, large tree diversity	Monodominant stands, less tree diversity	Low fruit abundance	Dipterocarps dominate, multiyear interval mass fruiting	Primitive conifers, dipterocarps
Canopy height	100–165 ft (30–50 m)	80–150 ft (24–45 m)	80–100 ft (24–30 m)	100–165 ft (30–50 m) with 230 ft (70 m) emergents	80–150 ft (24–45 m)
Distinctive animals	Small primates, large bird diversity	Ground-dwelling mammals, large primates	Lemurs, tenrecs, fossa	Orangutans, gliding animals, flying squirrels	Marsupials, monotremes, birds with elaborate mating displays, cassowaries

1,100 tributaries, 17 of which are longer than 1,000 mi (1,610 km), and two (the Negro and the Madeira) are larger than the Congo River in Africa in terms of volume. The Orinoco River is the second longest tropical river in South America and flows through Colombia and Venezuela and into the Atlantic Ocean. Tropical rainforests occur throughout its basin.

The Neotropical rainforest includes 45 percent of the total Tropical Rainforest Biome of the world. An estimated 1.08 million mi^2 (281.2 million ha) of these tropical forests remain. Underlying a large part of this area is one of the oldest rock formations in the world, the Precambrian shield. In the Neotropics, it is represented

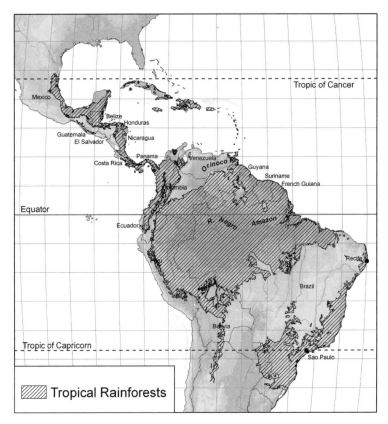

Figure 3.1 Location of tropical rainforests within the Neotropical region. *(Map by Bernd Kuennecke.)*

by the Brazilian or Guyanan shield. This ancient basement rock formation occurs in pieces in Africa and Asia as well.

The tropical rainforests of Central and South America reflect the history of the continent. Central America is a mix of geology of different ages and origins, whose biota before its connection to South America was dominated by North American species. Today, life in the rainforests of Central America is more similar to that of South America. South America was separated from Africa at the breakup of the Gondwana supercontinent and isolated for millions of years. Periodic connections to Central and North America caused by lower sea levels have contributed to the flora and fauna seen today. In addition, the development of the Andes mountain range continues to influence the region. Because or in spite of these events, the Neotropical rainforest contains a vast array of plants and animals found nowhere else in the world. The Amazon and Orinoco rivers and their tributaries throughout Brazil, southern Venezuela, eastern Colombia, Peru, and Bolivia reach more than a thousand miles of land, enriching the greatest rainforests in the world in terms of area and species diversity. The vast and lavish forest houses countless species, many that have evolved particular adaptations to exist in this moist, hot, and often-flooded environment.

The Amazon Basin contains a mosaic of ecosystems and vegetation types, including rainforests, seasonal forests, deciduous forests, and flooded forests, among others. The Amazon River and all its tributaries are the lifeline of these forests, and its history plays an important part in the development of rainforests. Earlier in history, when the continents were joined as part of Gondwana, the Amazon River flowed westward, perhaps as part of a larger Congo (Zaire) River system from the interior of present-day Africa. The South American continental plate separated from the African Plate and moved westward, crashing into the Nazca Plate. About 15 million years ago, the Andes were formed, forced up by the collision of these tectonic plates. The Amazon became a vast inland sea that gradually turned into a massive swamp. About 10 million years ago, through years of constant rainfall and heat, water worked through the sandstone and the Amazon began to flow eastward, permitting the development of the Amazon rainforest. During a series of climatic changes and Ice Ages, the tropical rainforests around the world expanded and contracted. Some researchers suggest that during drier times, much of the Amazon rainforest reverted to savanna and seasonal forest, leaving smaller isolated rainforests as refugia, areas of isolated populations. When the Ice Ages ended, the forest spread and these refugia joined, still maintaining some of their once-isolated divergent species. This may explain some of the incredible diversity within the Neotropical rainforest.

The Choco is one of the last coastal rainforest left in the world. The Choco rainforest extends along the Pacific Coast of Panama, Colombia, and northern Ecuador. It is recognized as a world hotspot for biodiversity and is home to a diverse assemblage of plants and animals. More than 11,000 species of plants, 900 species of birds, and at least 100 species of reptiles have been identified within the Choco with many of them endemic. The Choco receives more rainfall than the Amazon, totaling up to 630 in (16,000 mm) in some areas. The Choco was isolated from the Amazon with the formation of the Andes mountains. This isolation allowed for a rich array of unique species to evolve over time.

Climate

The climate of the Neotropical rainforest is consistently hot and humid. Temperatures average around 88° F (31° C) during the day, with nighttime lows around 72° F (22° C). Humidity is never less than 88 percent. Rain is abundant. In the Amazon Basin, precipitation ranges from 80–120 in (2,000–3,000 mm) annually. This precipitation comes from the Trade Winds and the Intertropical Convergence Zone (ITCZ), along with convective rainfall brought about by evapotransporation within the forest. Annual rainfall in the Choco can reach 157–315 in (4,000–8,000 mm).

The Neotropical rainforest is of major importance in regulating global climate. Carbon dioxide absorbed through photosynthesis acts as a sink, holding carbon dioxide in the rainforest biomass. When the forest is destroyed, the sink they once provided is destroyed.

Soils

Soils within the Neotropical rainforest are varied. They are classified into three main groups: oxisols, ultisols, and inceptisols. Oxisols make up about 50 percent of all Neotropical soils. They are deep red or yellow infertile soils. Two-thirds of all tropical oxisols are found in South America. They occur primarily in areas influenced by the Guyanan and Brazilian shield (remnants of the Precambrian basement rock). Oxisols are found throughout the Amazon and in Colombia along the Pacific Coast. The other deeply weathered soils in the Neotropics, the ultisols, occur in the Amazon Basin and along the eastern coast of Central America and Brazil. These tropical soils tend to have a high clay content, making them slippery when wet with high erosion potential. About 32 percent of Neotropical soils are ultisols.

The other soil types are inceptisols, entisols, and alfisols. Inceptisols occupy the largest portion of those soils. Half of the inceptisols occurring in the Neotropics occur on older alluvial plains along major rivers, others are of volcanic origin. Inceptisols are more fertile than oxisols or ultisols.

A unique white, sandy soil exists in parts of the Amazon Basin. This soil derives from the eroding and weathering of the ancient Brazilian and Guyanan shields and from former coastal beaches. These soils are extremely poor in mineral content and are infertile, having been weathered for hundreds of millions of years. They are extremely well drained and nutrient poor. A unique forest called a caatinga grows on these soils, but if it is removed, the forest does not return. The caatinga is described in further detail in Chapter 5, Regional Expressions of Tropical Seasonal Forests.

All soils of the Neotropical rainforest tend to be rather infertile and low in nutrients and minerals. Interestingly, recent studies document that sands from the Sahara travel across the Atlantic and enrich the soils of the Neotropical rainforest.

Vegetation in the Neotropical Rainforest

The Neotropical rainforest throughout its distribution remains similar in general plant structure and appearance to other rainforests (see Chapter 2). However, the particular plant and animal species within the forest are distinct. The vertical stratification of three canopy layers and two understory layers is typical. Along with these tree layers, are an accompanying array of woody vines, orchids, bromeliads, and other epiphytes (see Figure 3.2).

Forest Structure
Neotropical rainforests hold some of the tallest rainforest trees in the world. The emergents are widely spaced trees ranging on average from 100–165 ft (30–50 m)

Figure 3.2 Tropical rainforest cover large areas of Bolivia along the Madidi River. *(Photo by author.)*

in height. The tallest trees are found in the lowland tropical forests. Some emergents have reached heights of 300 ft (90 m), but because of continuous deforestation, these trees are increasingly rare. These trees contend with strong drying winds and tend to have small leaves. Some species may be semideciduous during the brief dry season. Many trees within the Neotropical rainforest have spreading, flattened crowns with branches radiating out from one or a few points like the spokes of an umbrella. This evolutionary adaptation provides the tree with maximum surface area exposed to the sun and limits the shading effect of the tree leaves on itself; lower in the tree canopy, the pattern is less pronounced. Neighboring trees competing for light often modify this round flattened-crown shape. The B layer is composed of a closed canopy of trees around 80 ft (25 m) tall. Light is readily available at the top of this layer, but greatly reduced below it. Woody vines and epiphytes are found among the B layer and can be the majority of biomass present in this layer. The C layer is composed of smaller trees around 60 ft (18 m) tall. This creates a closed canopy with little air movement and constantly high humidity below these layers.

The Neotropical rainforest is characterized by high floral and faunal species diversity. Along with tree diversity, there is an enormous diversity of nontree species of plants within the rainforest. Researchers studying forest understory have found that the number of vine species within the Neotropical rainforest is greater than in the rainforests of Africa and Asia. Many studies assessing the number of tree species within individual forested areas have found varying yet large numbers of individual species within the areas studied. Plant species estimates range from 105 per 2.5 acres (1 ha) in Puerto Rico to as many as 900 in 2.5 acres (1 ha) in the Ecuadorian rainforest. Increasing rainfall is positively related to high species diversity.

A number of tree families are represented in the emergent layer. Trees in the trumpet creeper (Bignoniaceae), Brazil nut (Lecythidaceae), and vochysia (Vochysiaceae) families occur in other forests but are most diverse in the Neotropical rainforest. Other plant families present among the canopy layers include the balsawood (Malvaceae), kapok tree (Bombacaceae), legume (Fabaceae), and fig (Moraceae) families, along with members of the euphorb (Euphorbiaceae), coffee (Rubiaceae), and palm (Arecaceae) families.

Tropical trees can have bark of a variety of colors from light to dark, with or without splotches, and in a variety of thicknesses. The wood is often hard and dense as a protection against the wood-eating termites that are plentiful. Tree bark is typically smooth to discourage the growth of vines on the tree. Some tree bark is particularly distinctive, such as the chicle tree (see Figure 3.3). The bark of the chicle tree is brown with gray spots, reasonably smooth, and deeply fissured. The sticky substance that lies behind the bark is extracted by cutting small gashes into the tree. Before the advent of a synthetic replacement, chicle was used in the making of chewing gum. Chicle is still used in some natural chewing gums.

Neotropical rainforest trees have broad buttresses at the base of the trunk. Leaves tend to be oval and have sharply pointed drip tips. Drip tips help facilitate rainfall drainage off the leaf and promote transpiration. Many leaves are thick and waxy, and remain on the tree for more than a year. Larger leaves are common among trees of the lower canopy layers. The large leaf surface helps intercept light in the sun-dappled lower strata of the forest. If these trees succeed in reaching the higher canopy layers, their newer leaves are often smaller.

What lies within the canopy of the Neotropical rainforest has been investigated only recently. Pioneering researchers, using new techniques for climbing and surveying the canopy, have been able to document new species and species interactions in and above the upper canopy. These researchers had to reach great heights and endure pelting rain, strong winds, swaying trees, and other untold hazards.

They found that the canopy is rich in flowering trees, orchids, bromeliads, vines, and other light-loving plants. Along with this flora abundance comes a diversity of insects, arthropods, amphibians, birds, bats, rodents, and other mammals that spend their entire life in the upper canopy, many of them never reaching the forest surface. Some canopy trees have been found to contain as many as 200 orchids, with 1,500 other epiphytic plants within their entire height. One layer of the canopy is in fact many layers of epiphytes, bromeliads, mosses, lichen, liverworts, and algae covering all types of structural surfaces, including trees, branches, woody vines, and even mammals, such as sloths.

Little light penetrates beyond the canopy layers. Shrubs are adapted to low light conditions and are common in the understory. They include members of families

Figure 3.3 This chicle tree (*Manilkara chicle*) was used in the production of chicle. Gashes are made into the tree to release the sticky sap. *(Photo by author.)*

found in the tree layers, such as the palm, legume, and coffee families, as well as those more restricted to the lower layers, such as the melastome (Melastomaceae), composite (Asteraceae), and pepper tree (Piperaceae) families. Heliconias, anthuriums, terrestrial bromeliads, and ferns are also found in the understory. Treefalls can open gaps, providing intense light to limited areas and allowing understory species that can take advantage of this sudden light to flourish temporarily. Seedlings and saplings of canopy trees can grow quickly in these gaps.

Flowers and Fruits

A visit to the Neotropical rainforest any time of year will yield a bright display of flowers and fruits. Flowers of tropical trees are colorful, fragrant, and often large. Due to the consistent day length and abundant rainfall, there is no one season of flowering. Tropical flowers come in an array of colors, sizes, shapes, and fragrances adapted to attract appropriate pollinators (see Plate V). Red, orange, and yellow flowers are associated with bird pollination. Hummingbirds pollinate a large number of Neotropical plants. Flowers that are light purple are usually insect pollinated. Some flowers pollinated by moths or bats flower at night, like the kapok tree. Fragrant flowers are attractive to bees, moths, and other insects. Many tropical trees pollinated by larger animals like bats and birds have large flowers, high in nectar content. Fruits can be small, medium, or large. Large fruits and seeds are more common in tropical areas. Palms such as the coconut produce large fruits with hard woody seeds inside. The monkey pot tree produces thick, large round fruits up to 8 in (200 mm) in diameter with up to 50 seeds. The Brazil nut tree produces large woody pods of seeds that open upon falling to the forest floor. Large fruits and seeds are a major food source for large tropical forest animals. Along with large birds, monkeys, rodents, peccaries, tapirs, and bats are common consumers of tropical fruits. They digest the fruit and aid in seed dispersal. In the flooded forests, fish are consumers and dispersers of tropical seeds. The majority of Neotropical rainforest trees have seeds dispersed by animals. Some trees produce smaller fruits or seeds that are consumed and dispersed by insects. Others trees like the kapok and mahogany have wind-dispersed seeds.

Many tropical fruits and trees can contain harmful substances to discourage predation. The seeds of the monkey pot tree are high in selenium. Some legume species have seeds in long pods containing toxic amino acids. Plants in the euphorb family contain a toxic, milky white substance; the sap from the para rubber tree is of economic importance in the production of natural latex. Some of these toxic substances have been discovered to have medicinal properties.

Some Neotropical trees exhibit cauliflory. Cacao, from which chocolate is derived, is cauliflorous. Several species of fig as well as others also exhibit cauliflory. Although insects pollinate some of the flowers of cauliflorous trees, larger bats and birds pollinate most of the flowers and consume the fruits.

Vines are an important structural feature within the Neotropical rainforest, and 130 plant families include some type of climbing species. Six hundred different vines have been identified in the rainforest, among them trumpet creepers (Bignoniaceae), squash (Curbitaceae), legumes (Fabaceae), milkweed (Asclepiadaceae), morning glory (Convolvulaceae), and catbrier (Smilaceae). Vines can be danglers, stranglers, or climbers. Lianas often dangle from the branches of canopy trees. Philodendrons are common bole climbers. Other climbers store nutrients in roots and tubers, like sweet potatoes. Figs (Moraceae) are common stranglers in the tropical forests around the world. About 150 species have been recorded in the Neotropics.

Epiphytes are abundant and play a significant ecological role in the Neotropical rainforest. More than 15,000 epiphyte species live in the forests of Central and South America. They consist of an amazing variety of lichens, mosses, liverworts, ferns, cacti, orchids, and bromeliads, to name only a few. Epiphytes in the pineapple (Bromeliaceae), cactus (Cactaceae), Panama hat (Cyclanthaceae), and black pepper (Piperaceae) families are almost exclusive to the Neotropical rainforest.

Orchids (family Orchidaceae) are commonly epiphytes here as well as in the other tropical rainforest regions. Some have bulbous stems that store water. Orchids use mycorrhizae to assist in their growth. Fungi living within the roots of the orchid aid in the assimilation of nutrients. In return, the orchid provides food and a home for the fungi. The relationship is considered mutualistic; that is, it is beneficial to both organisms. Other orchids are saprophytic and receive energy from decaying organic material.

Root Systems of Trees

Arial roots are common among tropical trees. They provide stability and aboveground aeration for plants in water-soaked soils. Walking palms (*Socratea* spp.) have stilt roots that seem to shift to take advantage of the available light (see Figure 3.4). The roots elevate the base of the trees often as much as a 3.2 ft (1 m) off the forest floor. These aerial roots of the walking palm allow the tree to shift position and "walk away" from hazards. As the tree sends out new roots, slowly moving away from its original point of germination, the lower trunk and older roots rot away and are left behind. The spines on the stilt roots are quite large and sharp to discourage climbing fauna and herbivory.

Homes in the Canopy

Within the Neotropical rainforest, epiphytic species of the pineapple family, Bromeliaceae, are the most common and abundant (see Plate VI). Bromeliad flowers grow from a central spike and are typically bright red, attracting the hummingbirds that pollinate them. Leaves of the bromeliads often overlap, creating a capture basin at their base. This small basin traps water and detritus and is home to a variety of organisms, including mosquitoes, spiders, snails, frogs, salamanders, and even crabs during some part of their life cycles. Bromeliads provide myriad benefits to canopy animals by providing pollen, nectar, and fruits to birds and mammals. They provide a source of drinking water for monkeys and other canopy animals. Bromeliads are Neotropical in origin and distribution. Only one species within the family occurs in tropical Africa. Some bromeliads grow in soil, but most grow on tree branches.

Flooded Forests

The Amazon River has an extremely shallow gradient, making it unable to transport all of the water that falls throughout the year. Because of this, water levels rise as high as 24 ft (7 m) in low-lying areas. During the rainy season, the forests of the Amazon Basin experience severe flooding. There are two types of seasonally flooded forests: the igapó forest and the várzea forest. The igapó forest grows along the black water rivers where the water is low in nutrients. These rivers have a black color caused by humic acid. The várzea forests grow along the white water rivers that carry fertile sediment from the Andes Mountains. The várzea forest has higher species diversity than the igapó forest. Trees in both forest types are adapted to annual flooding. They have thick, protective bark. If leaves are underwater, they stop metabolizing, but the tree does not shed its leaves. The trees continue to flower and carry fruits above the waterline. Many plants depend on this annual flooding to disperse their seeds. The rubber tree depends on fruit-eating fish to spread its seeds.

Figure 3.4 Walking palms are adapted to wet environments. *(Photo by author.)*

Animals of the Neotropical Rainforest

If it were not for the incredible plant diversity within the Neotropical rainforest, the diversity of animals would be far less. The complex structure, variety of habitats, and the abundance and diversity of leaves, flowers, fruits, and seeds leads to a vast array of animals. Animals and birds are more mobile than plants and can exploit conditions in different layers within the Neotropical rainforest. Most are adapted to live in specific parts of the forests. Few move between canopy and forest floor. Trees are the home for most birds, as well as monkeys, sloths, squirrels, rats and mice, frogs, snakes, lizards, and myriad invertebrates. Small stature, long limbs, and skin flaps, as well as other adaptations to move freely through the trees, are seen in birds, mammals, reptiles, and amphibians.

Other animals, such as ground birds, tapirs, anteaters, some snakes, and invertebrates spend their entire life on the forest floor. Larger animals, particularly herbivores, are rare because of limited food available on the forest floor, due to the rapid decomposition of fruits and restricted number of plants able to survival under the deep shade of the forest canopy. The fauna of the Neotropical rainforest today is the accumulative result of millions of years of geologic, climatic, and biological events.

Mammals

More mammals inhabit tropical rainforests than any other biome. Of the 500 species of mammals that have been identified in the Neotropics, 60 percent are endemic to the rainforests of South America. These include all platyrrhine monkeys, anteaters, opossums, echimyid rodents, and all living sloths. Some are highly specialized and adapted to life in the trees with prehensile tails, sharp claws, distinctive communication, and great leaping abilities. Others are generalists, feeding on a variety of food items and foraging in the trees, on the ground, and in the waterways.

Many of the mammals of the Neotropical forest have adapted to arboreal life by evolving prehensile tails and long limbs that allow them to swing from branch to branch and tree to tree within the forest. In addition, primates, porcupines, and sloths, among others, have opposable digits and long claws to hang on branches and trees. Binocular vision allows many to see clearly in the dense forest cover and have accurate depth perception, a clear asset when leaping from branch to branch. Other adaptations include varied feeding strategies or modifications in body parts to facilitate the use of the different resources available in the rainforest. For example, anteaters have developed long tongues that allow them to gather insects often hidden within logs. Sloths have jaws and teeth that allow them to chew leaves, their primary food sources. Porcupines in the tropical forest have jaws and teeth that allow them to consume tree bark.

The distribution of mammals is a combined result of geologic and biological history and evolution. By the late Cretaceous, early Paleocene (65 mya), South America along with Antarctica and Australia broke from Africa. As Antarctica continued to drift south and Australia drifted east, South America became isolated and remained that way for the next 50 million years. During this isolation, several endemic orders of mammals appeared on the South American continent, most of which are now extinct. Only the order Xenarthra (or Edentata), which contains anteaters, sloths, and armadillos, survives. Other Neotropical mammals—including the primates, marsupials, and rodents—have relatives on distant continents.

The uplift of the Andes Range was another significant event in the geologic history of the Neotropical rainforest. The rise of the mountains affected temperature, climate, and moisture regimes in high elevations, and presented an effective climatic as well as biological barrier to lowland forests. Another important event in Neotropical biogeography took place around 7 mya, when the Isthmus of Panama

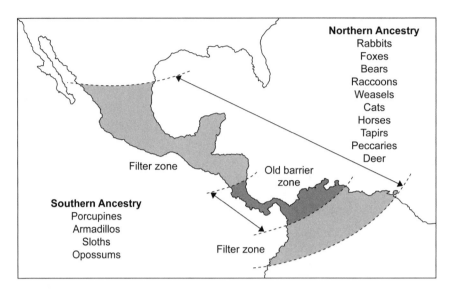

Figure 3.5 During the Great American Interchange, mammals and other organisms were able to move freely between Central and South America. *(Illustration by Jeff Dixon.)*

rose above sea level, allowing for a great interchange between South America and North America. While armadillos, caviomorph rodents (like capybaras and agoutis), primates, and marsupials moved south, squirrels, deer, peccaries, murid rodents, carnivores, and others moved north (see Figure 3.5). The exposure of this land bridge varied with fluctuating sea levels. The present fauna of the Neotropical rainforest is a product of ancestral taxonomic connections, temporary land bridges, recent uplift, and climatic changes. It includes families that evolved in both North and South America, and some whose origins may be Gondwanan.

The mammals families of the Neotropical rainforest include both marsupials and placentals (see Table 3.2). Marsupials are mammals with pouches to carry their young. All the marsupials in the Neotropical rainforest are opossums (family Didelphidae), a family restricted to the Americas, with one species in North America and about 70 in Central and South America. All 10 opossum genera are present in the Neotropical rainforest, including common opossums, woolly opossums, black-shouldered opossums, bushy-tailed opossums, short-tailed opossums, mouse opossums, water opossums, and four-eyed opossums.

Sloths spend most of their time in the middle and upper canopy, hanging upside down on branches or nestled into the fork of the tree eating leaves and shoots of trees (see Plate VII). They rest up to 20 hours a day.

Sloths have long, thick, coarse, brownish-grayish hair. The hairs are grooved longitudinally and—since the animal is usually hanging upside down—hang down from a center line on the belly (the opposite of other mammals on which the hair parts along the back). This allows sloths' fur to shed water while hanging. Under this

Table 3.2 Mammals Found in the Neotropical Rainforest

ORDER	FAMILY	COMMON NAMES
MARSUPIALS (METATHERIA)		
Didelphimorpha	Marmosidae	Mouse opossums
	Didelphidae	Opossums
PLACENTALS (EUTHERIA)		
Artiodactyla	Cervidae	Deer
	Tayassuidae	Peccaries
Carnivora	Canidae	Maned wolf, South American fox
	Felidae	Pumas, ocelots, and small cats
	Mustelidae	Weasels
	Procyonidae	Coatis, raccoons, ringtails
Cingulata	Dasypodidae	Armadillos
Chiroptera	Emballonuridae	Sheath tail bats
	Furipteridae	Smoky and thumbless bats
	Megadermatidae	False vampire bats
	Molossidae	Free-tailed bats
	Mormoopidae	Ghost-faced bats
	Natalidae	Funnel-eared bats
	Noctilionidae	Bulldog bats
	Phyllostomidae	New World leaf-nosed bats
	Thyropteridae	Disc-winged bats
	Vespertilionidae	Evening bats
Lagomorpha	Leporidae	Rabbits
Perrisodactyla	Tapiridae	Tapirs
Pilosa (edentates)	Myrecophagidae	Anteaters
	Bradypodidae	Three-toed sloths
Primates	Aotidae	Night monkeys
	Atelidae	Howler monkeys
	Cebidae	Capuchin monkeys
	Pitheciidae	Titis, sakis, uakaris
Rodentia	Agoutidae	Pacas
	Caviidae	Cavies
	Dasyproctidae	Agoutis
	Echimyidae	Spiny rats
	Erethizontidae	New World porcupines
	Hydrochaeridae	Capybara
	Muridae	Old World rats and mice
	Sciuridae	Squirrels
Soricomorpha	Soricidae	Shrews

The Last of the Edentates

The mammal group Xenarthra's origins can be traced back as far as the early Tertiary Period (60 mya), shortly after the decline of the dinosaurs. It is represented by three families: anteaters, sloths, and armadillos. They are the last remaining families of a large group of animals that evolved in South America when it was an isolated continent. With few exceptions, all members of these three families only occur in the Neotropics. All tend to be specialized feeders, eating mainly termites and ants, or rainforest canopy leaves.

Three genera and four species of anteaters are only found in the Neotropical rainforest. They are the giant anteater, collared anteater and tamandua (in the same genus), and the pygmy or silky anteater. Anteaters have no teeth, but a long tubelike snout and a sticky tongue that can be greatly extended to trap insects.

Two genera of the sloth family, the three-toed and two-toed sloths, exist solely in the Neotropical forests. Five species have been identified. The brown-throated and pale-throated three-toed sloths have a wide distribution in the Neotropical rainforest, while the maned three-toed sloth has a limited distribution along the Atlantic coast of Brazil. The two-toed sloths include Hoffman's two-toed sloth in Central and South America, with a disjunct population in southern Peru, and the southern two-toed sloth in South America. Sloths are found nowhere else in the world. The group Xenarthra also contains eight families of extinct ground sloths and armadillo-like mammals such as the giant sloth and glyptodont of the Pliocene Epoch.

coarse hair is a layer of thick fur, which protects the sloth from the bites of tropical ants that try to vigorously protect some of the trees sloths visit regularly, such as the cecropia tree. Sloths seems little affected by the ants. Three-toed sloths have long and sharp claws that are used to grip into the trees and to slash out at an enemy when threatened.

Although sloths are strictly arboreal, they will descend to the ground once a week to urinate and defecate, or to move to a more distant tree. On the ground, the sloth is vulnerable to predation from jaguars and anacondas. An even more formidable predator is the Harpy Eagle that locates a resting sloth while perched high in an emergent tree. The eagle swoops down and with its great talons easily and rapidly seizes the sloth.

In the rainy season, the sloth's hair has a greenish hue caused by microscopic blue-green algae that grow on its back. Along with algae, the sloth's hair is home to the pyralid moths whose adults live on their backs and feed on the algae growing there. Pyralid moths lay eggs in sloth dung where the larvae later hatch and feed. When the adult moth emerges, they fly up into the trees to find a new sloth host, and the cycle begins again. Other arthropods also live on sloths, including flies, mites, and beetles.

Sloths are endangered due to habitat destruction throughout the rainforests. Their populations are particularly vulnerable in Brazil, where forests are being destroyed at a rapid pace.

Several armadillos occur in the Neotropical rainforest, including the giant armadillo, the northern and southern naked-tailed armadillos, the yellow armadillo, and three different long-nosed armadillos. With the exception of the nine-banded long-nose armadillo, common in the southern United States, all are restricted to Central and South America. Like anteaters, armadillos specialize in ants, termites, and other forest insects. Larger armadillos have been hunted for meat, while smaller ones have been hunted for their shells, used in making a guitar-like string instrument in parts of South America.

Bats attain their greatest diversity in the Neotropical rainforest, where they are the most numerous and account for 39 percent of all mammals in the rainforest. All Neotropical bats belong to the suborder Microchiroptera. These bats use sonar or echolocation. The bats emit high-frequency sounds through their mouths or noses and the returning echoes give them information about their surroundings. They are mostly nocturnal.

Primates are another group of mammals commonly found in the Neotropics. Four Neotropical families form a distinct taxonomic group called the platyrrhines. These families consist of the New World monkeys (Cebidae), including marmosets, tamarins; squirrel and capuchin monkeys; night monkeys and owl monkeys (Aotidae); howler, woolly, spider, and woolly spider monkeys (Atelidae); and titi monkeys, sakis, and uakaris (Pithecidae). Platyrrhines are different from the catarrhine primates found in the tropical forests of Asia and Africa. Platyrrhines have short muzzles and flat, naked faces. Their nostrils are widely set and open to the side. Their eyes face forward, and they have short bodies, and long hind legs (see Figure 3.6). The origin of platyr-

Creatures of the Night

Walking through the forest at dusk, you will undoubtedly see mammals flying through the forest. Nine families of bats with 75 genera are present in the Neotropical rainforest. Of these, five families are endemic: the funnel-eared bats, thumbless bats, sucker-footed bats, bulldog bats, and leaf-nosed bats. Leaf-nosed bats include five subfamilies: the spear-nosed bats, long-tongued bats, short-tailed bats, fruit bats, and vampire bats. The four other families present are sheath-tailed bats, leaf-chinned bats, mustached and naked-backed bats, vespertilionid or brown bats, and free-tailed or Mastiff bats.

Bats occupy a vast array of niches. There are frugivores (fruit eaters), nectavores (nectar and pollen eaters), carnivores (meat eaters), insectivores (insect eaters), omnivores (generalists), and even sangivores (blood eaters). Bats are crucial in regulating insect populations, pollinating flowers, and dispersing seeds within the Neotropics.

rhines is a matter of considerable debate. Some researchers point to a primitive relative originating in Africa, with platyrrhines evolving from this relative in the Neotropics after South America and Africa separated. Others suggest a more recent origin that has primates coming from Africa through Asia and North America to South America when the northern landmasses had more tropical environments that facilitated migrations.

Most Neotropical monkeys have long tails and spend most if not all of their lives in trees. Many have prehensile tails (unlike their African counterparts). These tails assist in movement and stability. Primates are the most important seed dispersers for canopy trees and lianas, and with continuous forest habitat, many can travel over long distances.

Marmosets and tamarins are small primates with tufted hair, tassels, or manes on their heads; claws instead of nails; and long nonprehensile tails. Marmosets and tamarins feed on fruits and insects and live within the dense vegetation of the forest canopy. Squirrel and capuchin monkeys are medium-size diurnal, arboreal monkeys that eat fruit, leaves, insects, or small vertebrates. Squirrel monkeys do not have prehensile tails, but capuchin monkeys do. Night or owl monkeys are the only family of nocturnal primates in the Americas; they feed on fruits, insects,

Figure 3.6 Squirrel monkeys can be found in the Neotropical rainforests, but their populations are dwindling due to deforestation, as shown here in Bolivia. *(Photo courtesy of Gareth Bennett.)*

and flower nectar. Night monkeys are common and often seen near human settlement.

Titi monkeys, sakis, and uakaris are a group of specialized diurnal, medium-size, arboreal monkeys. They have bushy nonprehensile tails that hang down from tree branches and are often visible from below. This group is restricted to the Amazon Basin and the Guyanas. A new species of titi monkey, the Golden Palace monkey, was discovered in 2005 along the border of Madidi National Park in Bolivia. The winner of an Internet auction that raised funds for conservation named the monkey.

Spider, woolly, woolly spider, and howler monkeys are large monkeys with well-developed prehensile tails. Spider and woolly monkeys typically swing from their arms or tails. The spider monkey has elongated limbs and tail, and is the most acrobatic of the group. It can leap as far as 30 ft (9–10 m). Howler monkeys are slow or even sedentary. They have developed elaborate ways of communication within the forest canopy, and their voices carry for at least 10 mi (16 km) through the forest at dawn and at dusk. These large monkeys of the Neotropical rainforest are hunted intensively for their meat, and most are rare or endangered. Much of their habitat is now gone or threatened by deforestation and continued fragmentation, limiting their ability to travel in search of food or mates.

Tapirs are the only odd-toed ungulates (order Perissodactyla) in the Neotropical rainforest. Worldwide the family Tapiridae consists of just one genus with four species. Three of these species are Neotropical and are the only extant native odd-toed ungulates in the New World. Weighing up to 670 lbs (300 kg), the tapir is the largest land mammal in the Neotropical rainforest. Tapirs are herbivores and need large amounts of plant matter every day. Their noses are prehensile, allowing them to grab onto and pull up plants. Both the Brazilian tapir and Baird's tapir occur in the rainforest. The Brazilian tapir's range includes South America east of the Andes from northern Colombia to southern Brazil, Paraguay, and northern Argentina. Baird's tapirs occur further north in Central and South America from southern Mexico through Panama and west of the Andes through northern Colombia and Ecuador. (The third species is the mountain tapir. It is the rarest and occurs only in high elevations in the equatorial Andes, Colombia, Ecuador, and Peru.)

Two families, peccaries and deer, represent the even-toed ungulates (order Artiodactyla). Peccaries are medium-size pig-like animals with large heads and pig-like snouts, thick necks, and large stocky bodies with practically no tail. They are diurnal and feed on fruit, nuts, vegetation, snails, and other small animals. All are restricted to the Americas. The collared peccary and the white-lipped peccary are found in the rainforests of Central and South America into Argentina. (The collared peccary has moved into drier areas in the United States in Texas and Arizona.) The third species is the rare Chacoan peccary, which lives in the Gran Chaco of northern Argentina, southeastern Bolivia, and western Paraguay (see Chapter 5).

Representatives of the deer family in the Neotropical rainforest are the red brocket deer, whose range extends from Central America into South America, and the brown or gray brocket deer, whose distribution is restricted from Panama into South America, east of the Andes. Rainforest deer are rather small and solitary, diurnal, and feed on leaves, fallen fruits, and flowers.

Rodents (order Rodentia) are widespread, varied, and by far the most diverse group of mammals worldwide. Three major groups of rodents occur in the Neotropics: squirrels, mice and rats, and caviomorph or hystricognath rodents, among which are the world's largest rodents.

Squirrels are highly variable in coat color and pattern geographically. Eighteen species have been identified in the Neotropical rainforest. All are diurnal and arboreal, with some foraging in the trees and on the ground. They typically feed on fruits, flowers, nuts, bark, fungi, and some insects. Many have restricted distributions.

Grass mice, pocket mice, Mexican deer mice, long-nose mice, house mice, harvest mice and spiny mice, rice rats, forest rats, water rats, armored rats, climbing rats, crab-eating rats, spiny rats, bamboo rats, and tree rats are just a few of the native mice and rats found in the Neotropical rainforest. They occupy a variety of niches, feeding on myriad fruits, plants, fungi, and invertebrates. Some are aquatic, some terrestrial and some spend their entire lives in trees. Many are restricted in distribution, but others are common throughout the rainforests of Central and South America.

The largest Neotropical rodent group includes the porcupines, and the true cavy-like mammals: agoutis, pacas, acouchys, pacaranas, and the largest of all rodents, the capybara. Most porcupines are nocturnal and arboreal, while the cavy-like mammals are terrestrial and mostly diurnal. Agoutis, pacas, acouchys, and pacaranas forage during the day or night. The semi-aquatic capybaras are naturally diurnal, but they become nocturnal in areas where they are heavily hunted.

Only one rabbit occurs in the Neotropical rainforest, the Brazilian rabbit or tapiti. It is nocturnal, terrestrial, and solitary and feeds on grass and small plants.

Carnivores play an important part in regulating the populations of prey species. Members of the dog, raccoon, weasel, and cat families comprise the carnivores of Neotropical rainforest. Tropical carnivores tend to be omnivorous, feeding on insects, fruits, and leaves in addition to animal meat. Most are opportunistic and do not specialize on a particular prey item.

Little is known about the bush dogs of Central and South America and the short-eared dogs of South America, but both are thought to be diurnal. Short-eared dogs probably hunt alone, while bush dogs often hunt in small groups of two to four individuals.

The raccoon family includes two species of raccoons, two species of coatis (coatimundis), the olingo, the kinkajou, and the cacomistle. Typically nocturnal, these arboreal animals feed on fruits, invertebrates, and small animals both on the forest floor and in the canopy. All raise their young in tree nests. The northern and crab-eating raccoons are typically nocturnal and terrestrial; both are good climbers, escaping to the trees for protection from predators. The two coati species are restricted in distribution. The range of the white-nosed coati begins in Texas and Arizona in North America and extends through Central America and along the coasts of Colombia, Ecuador, and Peru to the most northwestern tip of South America. The South American coati is found east of the Andes throughout Colombia and Venezuela south to Argentina and Uruguay. Coatis are both terrestrial and arboreal, and tend to be diurnal, feeding on fruits and invertebrates (see Figure 3.7). Olingos, kinkajous, and cacomistles are strictly arboreal, nocturnal, and solitary animals. Their diet is primarily fruit, with occasional insects, flowers, and small vertebrates.

Members of the weasel or mustelid family include weasels, grison, skunk, hog-nosed skunk, tayra, river otter, and giant otter. Although some are good climbers, all forage on the ground or in water. Mustelids have dense fur and an extremely powerful bite that enables them to kill prey larger than themselves. They feed on rodents, birds, insects, fruit, fish, crustaceans, snakes, and small caimans. They can be diurnal, nocturnal, or crepuscular (hunting at dawn and dusk). Because their fur can be valuable, they are often hunted. The Amazon weasel, the rarest carnivore in South America, is restricted to lowland forests east of the Andes in Peru, Ecuador, and Brazil. The grison and tayra occur in both Central and South America, as does the southern river otter. Giant otters are restricted to South America north of Argentina.

Figure 3.7 Coatis are commonly seen in the rainforests of Costa Rica. *(Photo courtesy of David B. Smith, Ph.D.)*

Excluding humans, cats are the main predators in tropical rainforests all over the world. They have teeth specialized for killing and eating meat. Most hunt alone searching for small mammals, snakes, turtles, caimans, birds, fish, and insects. Most rainforest cats are nocturnal. Seven occur in the Neotropical rainforest: the smaller jaguarundi, oncilla, margay, and ocelot; and the larger puma, panther, and jaguar. All range from Central America throughout South America, although the distribution of the oncilla, a house cat-size species, is poorly known.

Neotropical Birds

Neotropical bird life is unequaled in any other part of the world. More than 1,400 species have been identified within the rainforest, with almost a third endemic to the lowland evergreen forest.

Birds of the forest have evolved bright coloration. This allows them to blend in with the dappled light of the upper canopy as well as the many colorful flowers and fruits high in the forest. Parrots, macaws, toucans, and hummingbirds, to name a few, produce vibrant colors in the canopy trees and the air above.

In the lower canopy, heat, moisture, evergreen vegetation, and abundant food have allowed some tropical birds to develop a more sedentary lifestyle. Short, broad wings allow birds to maneuver through the trees. Without the need to travel long distances for food or shelter, resident birds of the understory have less need for the slender pointed wings of some of their seasonal companions from the mid-latitudes that winter in the rainforest.

Many birds have feeding specializations that take advantage of the vast array of potential food sources within the Neotropical rainforest. Large insectivorous birds endemic to the Neotropical rainforest include potoos, puffbirds, motmots, and jacamars. Smaller endemic insect feeders include antbirds, ovenbirds, wrens, and vireos, to name just a few. These birds exploit all the layers of the forest from the ground to the understory and upper canopy, as well as different plants or parts of plants (bark, twigs, leaf undersides, and so on).

Some birds, including many parrots, macaws, and parakeets, feed primarily on fruit (frugivores). The cave-nesting Oilbird specializes in palm nuts. Toucans, hoatzins, and trogons feed primarily on fruits and seeds, but also take insects on occasion. Many of the frugivores are separated by size as well as where they get their food. Toucans, parrots, and macaws pick fruit from canopy trees; smaller birds like manakins and tanagers exploit fruit, flowers, and seeds in the lower canopy layers. Quetzals are beautiful but weak fliers; they are fruit eaters in the understory and lower canopy. Curassows and pigeons take advantage of fruit that has fallen to the forest floor. Beaks on birds like parrots and macaws are hooked and powerful and are able to pierce the tough tusks and skins of fruit to uncover the juicy pulp and seeds within. Other birds have developed beaks suitable for prying seeds from hard-coated fruits.

Hummingbirds are restricted to the Americas. They are nectar feeders and different species evolved diverse bill lengths that allow each to specialize on a different set of flowers. Many tropical hummingbirds have curved beaks.

The Neotropical rainforest hosts a variety of ground birds, including trumpeters and terrestrial cuckoos, curassows, guans, chachalacas, and tinamous, as well as pigeons. These birds consume small reptiles and insects, as well as fruit that has fallen to the forest floor.

The rainforest is home to a variety of carnivorous birds. Diurnal raptors vary in size from the Tiny Hawk of Panama weighing 2.5 oz (75 g) to the majestic Harpy Eagle that can weigh more than 20 lbs (9 kg). The rainforest also hosts a diversity of falcons and caracaras that hunt along the many waterways. Owls are nocturnal hunters within the forest. Vultures are important carrion feeders in the Neotropics.

Reptiles and Amphibians

Snakes, turtles, lizards, crocodiles, and frogs are among the many reptiles and amphibians that inhabit the Neotropical rainforest. Snakes are grouped as vipers and pit vipers, cobras and coral snakes, constrictors (boas and pythons), and other nonconstricting, nonpoisonous snakes. Pit vipers are the most deadly snakes in the world. Pit vipers range from North America into the Neotropics and throughout South America. Pit vipers have large triangular shaped heads and slit shaped pupils (see Plate VIII). They are named for the sensory depression or pit located between their nostrils and eyes that they use to sense warm-blooded prey. Pit vipers have sharp needle-like fangs that can deliver a lethal dose of poison and affect the blood tissue or nervous system of their prey. They mostly eat small mammals and birds. A number of nonvenomous snakes in the Neotropical rainforest have taken on the behavior of pit vipers, coiling, striking, and even flattening the head to resemble the distinctive triangular shape of a pit viper. This type of mimicry attempts to take advantage of the threat posed by the truly dangerous snakes by imitating their appearance and gestures.

Coral snakes are another type of poisonous snake found in the Neotropics. Coral snakes are best recognized by their coloration of red, yellow, and black bands. Coral snakes have short fangs that inject venom that directly affects the central nervous system, producing paralysis and death by suffocation. Coral snakes are related to the cobras and mambas of Asia and Africa. They are typically 2–4 ft (600–1,200 mm) in length and eat mostly lizards and other snakes. Several less harmful or nonpoisonous snakes take on the coloration of coral snakes to warn off potential predators. These look-a-likes are often called false coral snakes.

The other large group of snakes within the Neotropical rainforest and other tropical forest regions are the constrictors (Boidae), the world's largest snakes. In the Neotropics, these snakes are called boas; in Africa and Asia, they are called pythons. These snakes are not poisonous, but have sharp teeth. Characterized by long wide heads with pointed snouts and wide bodies, they capture their prey by attacking and biting it, coiling around it, and tightening their grip until they suffocate the animal that they swallow whole. The most common and well-known boa in the Neotropical rainforest include the beautiful, deep-green-colored emerald tree boa and the smaller rainbow boas, both excellent tree climbers. The anaconda, the largest of the Neotropical constrictors, reaches nearly 30 ft (10 m) in length and lives in and along swamps and rivers feeding on large rodents, peccaries, large birds, tapirs, and even crocodiles. It rivals the rock python of Africa and reticulated python of Asia in size and weight.

Pit Vipers of the Neotropics

The Neotropical rainforest is home to the largest pit viper in the world, the bushmaster, which reaches lengths of 6.5–11.7 ft (2.5–3.0 m). Bushmasters inhabit the lowland forests of Central America and occur throughout the Amazon forest and the coastal forests of southeastern Brazil. One of the most notorious of pit vipers in the Neotropics is the fer-de-lance. It is one of 31 lancehead vipers that range from Central America into the Orinoco Basin of South America. Fer-de-lances are found in lowland areas. One snake can typically produce 50 or more offspring in one season. Recent climate warming in Central America now allows the fer-de-lance to reproduce twice a year. This creates a huge population of snakes that have been leaving the forest and entering plantations in search of food, thereby endangering the lives of plantation workers and their families. Pit vipers are responsible for more human fatalities than any other snake in the Neotropics. In humans, their venom is fast acting and causes the rapid destruction of blood cells, producing an infection that results in severe necrosis of the bite site and the surrounding tissue. When antivenom is provided rapidly and the wound is properly treated, mortality from a bite is low. In addition to bushmasters and fer-de-lances, rainforest pit vipers include forest pit vipers, palm pit vipers (also called eyelash pit vipers), and hognose pit vipers. (Rattlesnakes, occurring through North and Central America, are also pit vipers.)

In addition to the pit vipers, coral snakes, constrictors, and their imitators, vine snakes and other snakes spend their lives in and about the canopy, feeding on small birds, lizards, and small frogs.

Other reptiles in the Neotropical rainforest include lizards, geckos, and tortoises and turtles. Iguanas are common larger lizards in the tropics common. They can be greenish-brown when young, but become darker brown as adults. They feed on insects when young and mostly fruits and leaves when adult. Iguanas can be spotted basking in the sun in the trees and along the shores of rivers. Black iguanas (ctenosaurs) are large lizards in the forest. They burrow in the ground as well as climb trees and eat mostly fruit and plants, but will take baby birds, eggs, and bats on occasion. Anoles are small green or brown lizards. Some can change their color to blend into the background. Anoles account for 50 percent of all lizards in the iguana family. Their toe pads make them effective tree climbers, and they can be found perched on tree branches at great heights or on tree trunks and the forest floor. Anoles eat mostly insects.

The basilisk lizard is also known as the "Jesus lizard" due to its ability to seemingly walk on water. Basilisks are large green or brown lizards with prominent fin-like crests, dewlaps (hanging skin beneath the jaw), and powerful legs. They can run on their two hind legs short distances across the surface of water. Their toes have a fringe of scales that provide them with a larger surface area to accomplish this feat. Their tails help them balance. Basilisks feed on vertebrates and invertebrates, as well as fruits and flowers. They are primarily found in the tropical rainforests of Central America.

Tegus are the largest South American lizards. Tegus have large bodies; the common tegu can reach lengths of 55 in (140 cm). Tegus range throughout tropical and subtropical South America, but occur nowhere else. They live in the forest and forest edges eating small animals, eggs, and the occasional chicken. Tegus are thought to be the most exploited reptile in the world and are hunted for local food as well as for the international pet and skins trade.

Geckos are a diverse and wide-ranging family of lizards. They are small to average in size and are found in warm climates around the world. Most geckos have no eyelids and instead have a transparent membrane that they lick to clean. Geckos have suction-like scales, called lamellae, on the undersides of their feet. They are nocturnal and hunt for arthropods. Their loud chirping sounds can fill the forest night.

Turtles in the Neotropics are primarily side-necked freshwater turtles. Rather than pulling their heads directly back under their shell, they tuck them in sideways. Turtles were common along the Amazon before excessive hunting and egg collection seriously reduced their populations. They are found on logs and alongside waterways.

Alligators, caiman, and crocodiles are commonly seen along riverbanks within the Neotropical rainforest. The four Neotropical species of crocodiles are mostly confined to the coastal mangroves and riverine habitats of Belize, Costa Rica,

Cuba, and Colombia. Alligators and caimans are more common; but because of human pressures, their populations have decreased significantly. Caimans can be found from southern Mexico through northern Argentina. Most caiman species grow to lengths of about 8 ft (2.5 m). The black caiman can reach lengths of nearly 20 ft (6 m). The black caiman, along with several other caiman and Neotropical crocodiles, are endangered.

Amphibians of the Neotropics belong to three orders: salamanders and newts, caecilians, and the largest group, frogs and toads. All amphibians require water to reproduce and typically lay eggs in a gelatinous sac in ponds or streams, but in the tropical rainforest, some lay eggs in bromeliads high in the canopy. Adults have semipermeable skin and usually need to reside in moist environments; toads are the exception.

Frogs are the most abundant and diverse amphibians in the Neotropical rainforest. Of the approximately 4,300 species of frogs and toads identified worldwide, 1,600 occur in the Neotropical rainforest. Frogs occupy a variety of niches and have evolved strikingly different reproductive strategies. These include live births of fully formed frogs, eggs laid on plants where the tadpoles can hatch and fall into the water, eggs laid in foam nests in bromeliads and in tree cavities, and even eggs that upon fertilization are moved to the mother's back, where the growing tadpoles remain until they develop into fully formed frogs.

Pipid frogs are common and have flattened bodies and are tongueless. All are aquatic with webbed feet. Frogs in the genus *Pipa* live in eastern Panama and South America east of the Andes. The Surinam toad is about 7 in (178 mm) long with a pointed head. It has a peculiar reproductive adaptation in that the male fertilizes the eggs and distributes them over the female's back, where they become embedded into small pockets and develop, emerging as tiny frogs.

The Neotropical frogs inhabit trees and shrubs, as well as leaf litter on the forest floor. Most have direct developing eggs; that is, the tadpole stage is skipped and a fully formed frog emerges from the egg. Many leptodactylid frogs lay their eggs in foam nests that float on open water or deposit them in a tree cavity, burrow, or other sheltered site. The foam provides a protective coating for the eggs and tadpoles.

Toads are also common. *Bufo* species tend to be short legged and have heavy bodies, wart-like glands on the body and legs, and a round or oval parotoid gland behind the eye. They are toothless and tend to hunt at night. *Atelopus* is a genus occurring in Central and South America made up of diurnal toads that are brightly colored and produce a toxin. The coloration is thought to serve as a warning mechanism for potential predators.

Tree frogs are typically arboreal and possess well-developed finger and toe pads to aid in climbing. Their eyes are positioned to provide binocular vision ahead and at the periphery. Tree frogs of the genus *Phyllomedusa* have the first finger opposable to the other three, allowing them to grasp twigs and stems. Some tree frogs lay their eggs on plants above water so that the tadpoles can drop into the water when they hatch. Others lay their eggs in water-filled holes in trees or bromeliads. Several genera brood their eggs on the female's back.

Glass frogs occur in tropical forests from Mexico to Bolivia. Their abdominal skin is scarcely pigmented, making their intestines visible. They tend to be small and spend their lives in the forest trees.

Another family of frogs with a Neotropical distribution is that of the poison dart frogs. One group is dull-colored, nontoxic, and lives alongside rivers and steams. The second group consists of the poison frogs. These usually small to tiny colorful frogs tend to be diurnal. They come in bright blue, red, yellow, black, green and black, and other color combinations (see Plate IX). Best known are the poison dart frogs, whose skin excretes an alkaloid poison that affects the nervous system of those who touch it. This acts as a defense mechanism against would-be predators. Indigenous tribes of the Choco in Colombia rub the poison on the tips of their blowgun darts when preparing to hunt. The toxicity of poison dart frogs varies from species to species. Their bright colors serve as warning to any would be predators. These frogs will often lay their eggs in humid places high in the canopy, where they are typically guarded by the male or female. When the tadpoles are formed, they crawl onto their guardian's back and are carried to water.

The microhylid frogs occur throughout the tropical forests of the world. They include both arboreal and burrowing frogs. Three other South America families of frogs—Ruthven's frog, gold frogs and tree toads, and paradox frogs—each with very few members, all occur in the rainforest. The paradox frog is named because its tadpoles grow to remarkable sizes, up to 10 in (250 mm) in length, yet after metamorphosis, the largest adult frog is only 2.5 in (70 mm).

Salamanders are not as numerous as frogs in the Neotropics, but the lungless salamander family is quite diverse. These small, slender amphibians live in the deep shade of the forest. They are either fossorial or arboreal.

Caecilians are legless, tailless amphibians that are numerous in South America and found in the tropical rainforests throughout the world. Caecilians resemble earthworms and move through underground tunnels.

Fish

Fish play important roles in flooded forests of the Amazon and its tributaries. About 200 different kinds of fish consume the seeds and fruits of forest trees and are active dispersing agents. Seasonal flooding of the rivers lets fish swim among the trees and forage on dropped fruits. Some tropical fish also eat woody plant material, leaves, and detritus as well as invertebrates, small vertebrates, and zooplankton.

Insects and Other Invertebrates

The Neotropical rainforest is home to innumerable insects and other invertebrates that exploit whatever opportunities they can. Invertebrates evolved more than 350 mya and are quite adept at adaptation and survival. They have endured countless geologic and climatic changes and are extremely resilient to environmental changes. They play a vital role in the maintenance of the tropical rainforest as pollinators and decomposers, and they provide essential nutrition to myriad animals that inhabit the forest. The actual number of insects within the tropical rainforest is unknown. Several studies have documented thousands of species present on a single tree. By extrapolating these numbers to the larger forest, estimates of hundreds of thousands of species are thought to inhabit the rainforest. Although not all of the species identified within the Neotropical rainforest can be addressed in detail in this volume, a few groups will be examined.

Ants (order Hymenoptera) play a crucial role in destroying and recycling vegetation; others are carnivorous and able to strip a small animal in minutes. Leaf-cutter ants are a subgroup of the fungus-growing ants (Formicidae) present only in the Neotropics. They are thought to have evolved in the wooded and savanna areas of South America and to have diversified to take advantage of the lush vegetation of the rainforest. Large, strong worker ants climb selected trees, cut out small sections of leaves or flowers with their sharp mandibles, and carry them great distances to their nests. They use the leaves as organic mulch to nurture the growth of mycelial fungus—their sole source of food.

Army ants are swarming ants that are most abundant in the Neotropics and Africa. Army ants are social ants with a queen, soldiers, and worker classes. They form large colonies. Populations within one group can reach in excess of 1 million individuals. They can be orange to dark red, brown, or black. Neotropical army ants are nearly blind, and they communicate by chemical signals. Nomadic, they only stop at underground nests or in hollow logs during their reproductive cycles. Typically, one queen stays at the nest until the next move. Army ants practice group predation—the pack swarms their prey, killing and dismembering it to bring it back to their temporary nest. Army ants hunt on the ground and in trees. Prey includes caterpillars, spiders, millipedes, small frogs, lizards, salamanders, snakes, and small birds. Army ants conduct mass raids during which they remove large amounts of food. A single colony can harvest up to 90,000 insects during one day.

Bullet ants are large tropical ants. They can be as large as 1 in (3 cm) long. Their range includes Central and South America. They live at the base of trees or in tree cavities. Their attack mode is a hard bite delivered to the prey, accompanied by a powerful sting. The sting carries a neurotoxin that affects the nervous system of the prey. Bullet ants tend to be solitary when hunting and will also collect nectar, water, and plant parts, as well as arthropods, insects, and parts of small vertebrates.

Other ants have developed symbiotic relations with Neotropical plants. The cecropia tree has leaves and fruit that many creatures find desirable, but these tasty products are protected by Aztec ants. The tree's hollow trunk and branches provide a home for the ants, and the tree produces Mullerian bodies, glandular nodules rich in carbohydrates that the ants use as food. Ants consume these energy-rich capsules and viciously attack anything that touches the plant. Only the sloth, protected by an undercoat of dense fur, may safely climb the tree and dine on the leaves.

Various acacias have evolved similar symbiotic relationships with other ant species in other forests of the world.

Termites (order Isoptera) play a main role in decomposition and recycling within the Neotropical rainforest. Termites are social animals living in large groups. They nests in tree cavities, stumps, or on the soil surface and are a common sight in tropical rainforests around the world. Termites have a class structure in which individual roles are clearly defined. The colony comprises workers, soldiers, and the queen. Termites digest wood and forest litter with the help of a complex intestinal community of protozoa. In this mutualistic relationship, the termites provide the single-celled organisms with food and shelter and the protozoa allow the termite to process vast quantities of wood. Some researchers hypothesize that termites contribute to global warming since their vast populations produce large amounts of the greenhouse gases methane and carbon dioxide as byproducts of their digestive processes.

Neotropical butterflies and moths (order Lepidoptera) are highly diverse. A few families, most genera, and practically all species are endemic to the Neotropical region. Butterflies in the Neotropics include the brightly colored swallowtails, the whites, and the blues. The whites can be white, yellow, or orange; small to medium in size; and often present in large numbers along riverbanks. The blues are very small butterflies with diverse natural histories. They feed on fruits, nectar, and even animal carcasses. One, the ant butterfly, has a specific relationship with ants. The females tend to cluster around roaming army ant hordes, feeding on the droppings of antbirds that are also following the ant group.

The most prominent butterflies of the Neotropics are the nymphs. A large number of species have been identified in the rainforest, with as many as 1,800 described for one area alone. Nymphs are a highly diverse group of butterflies. The largest is the owl butterfly, which has a large eye-shaped marking on both wings. Only visible when the wings are open, the "eye" serves to distract potential predators. Another nymph, the blue morpho, is also large; it is one of the most brilliantly colored butterflies in the rainforests of Central and South America. The males have bright, iridescent blue upper wings with rather cryptic undersides. The blue morpho has a wingspan of more than 6 in (15 cm). Many tropical butterflies feed on rotting fruit and nectar. Many nontoxic butterflies of the Neotropics mimic toxic counterparts.

Less is known about rainforest moths. Caterpillars of tropical moths are often plant eaters, but some are leaf miners, stem borers, flower feeders, and fruit and seed eaters. The pyralid moth lives within the fur of sloths and lays its eggs in sloth dung when the animals come down from the trees to defecate.

The forest hosts a vast array of beetles (order Coleoptera). Some are brightly colored, others nondescript. Some are large, like the rhinoceros beetle, 3 in (8 cm) in length, and recognized by its long upcurved horn-like projection (see Figure 3.8). Others are very small. Some common beetles include horned beetles, dung and carrion beetles, harlequin beetles, fungus beetle, and wood-boring metallic beetles. Cockroaches (order Blattodea) are common decomposer in the forest.

Figure 3.8 Rhinoceros beetles can be found throughout the tropics. This one is feasting on the core of a pineapple. *(Photo courtesy of Jacob Holzman Smith.)*

Mosquitoes and other disease-carrying insects tend to feed on canopy animals, and few descend to the forest floor. Some may carry diseases such as malaria, yellow fever, and dengue fever among others. Once the forest is cleared, disease-carrying mosquitoes can become a major problem.

Spiders, scorpions, and centipedes are abundant in the Neotropical rainforest. Wolf spiders have small bodies and long legs; tarantulas have larger bodies. A particularly large tarantula, the South American goliath, is known to eat birds. Its leg span can be 7 in (18 cm). It has been observed capturing small birds, as well as small reptiles. Orb spiders build webs of such strong silk that it has been used in the production of some safety clothing. The ant spider impersonates an ant by exhibiting ant-like movements and holding its front legs forward to resemble ant antenna. This provides an effective camouflage to prey on its preferred diet, ants. Social spiders communally build large webs to trap prey.

Scorpions are also present, but rather than bite, they sting. Their sting can be toxic and irritating, but it is rarely fatal to large vertebrates. Centipedes and millipedes are both common forest creatures. Centipedes are nocturnal predators that feed on other invertebrates. They have powerful jaws that inflict a painful bite. They also use poison to subdue their prey. Millipedes are generally active during the day and feed on soft decomposing plant matter. Some species can reach up to 10 in (27 cm) in length.

The great diversity of plants and animals in the Neotropical rainforest described above merely provides a glimpse into the complex equatorial biome. Although species numbers are high, populations tend to be small or isolated. Limited population sizes contribute to the fragility of the forest. The destruction of one area of rainforest can lead to the extinction of hundreds of plants and animals that occur only in that region.

Human Impact

Much of Central America and many of the Caribbean Islands were once forested with tropical rainforest. Today, few Caribbean islands retain any primary forest cover, while fragments of rainforest continue to persist in some parks and reserves. In Central and South America, rainforest are being lost or rapidly degraded. The huge Amazon forest is largely intact, but it is rapidly being cleared at its margins and along an increasing network of roads. The Choco is similarly at risk. Estimates of rainforest loss ranges from 15 to almost 30 percent of the original forest. In the Amazon, around 11 million ac (4.3 million ha) were cleared per year between 2000 and 2005, and deforestation rates continue to increase. Two of the main stimuli for rainforest destruction in the Neotropics are large-scale ranching and slash-and-burn farming.

Large-scale ranching is usually initiated by bulldozing an area, extracting the valuable timber, and then burning the forest. Within the Amazon, an estimated 70–75 percent of deforestation is caused by large- and medium-scale ranching operations. This is true in other parts of the Neotropical rainforest as well. Slash-and-burn farming practices are typically conducted by small landowners. One to two acres (<1 ha) of forest is cleared, the understory is cut down with machetes and the debris is burned. The ash provides a pulse of nutrients available temporarily for crops. After a few years, the farmer moves on to a new area to repeat the process. With increasing populations and increasing demands on the land, large areas are affected by this short-term production strategy. Government programs encourage settlement in the forests by creating roads and providing free land for would-be farmers and thus exacerbate the problem. Modern industrial agriculture is also rapidly threatening the rainforest. Large tracks of land are being cleared for the production of export crops such as soybeans, palm oil, coffee, pineapples, bananas, and vegetables.

Industrial timbering has increased dramatically in the Neotropics. Most logging is done selectively, but clear-cutting also occurs. The indirect effects of logging come from the creation of roads, tracks, and clearings; soil erosion; an increase in invasive species; microclimatic changes; and tree mortality in surrounding areas. Multinational timber companies from Asia are active, buying up large tracts of land or obtaining long-term leases. Asian multinational corporations control at least 50,200 mi^2 (13 million ha) of the Amazon forest. Illegal logging also threatens the forest.

Wildfires are increasing in Neotropical rainforests. Most are human caused, as the undisturbed rainforest is not prone to fire. Fires are so prolific that they can be seen easily on nighttime satellite photography of the region. The forest recovers slowly. Increased fires alter the microclimate because the forest is no longer capable of maintaining an internal water balance. This can lead to regional climate changes, such as decreased precipitation, which make the forest even more susceptible to fire. Other localized impacts include gold mining and the consequent poisoning of waterways related to unsustainable mining practices. This threatens the health and lifestyles of indigenous people, as well as the forest. Oil and natural gas developments also have been increasing.

Forest destruction leads to widespread fragmentation of habitat, which causes such ecological damage as species loss, interruption of ecological processes, disruption of pollination, loss of carbon storage, and reduction of nutrient recycling. As human populations continue to grow and increase the demand for land and resources, continued destruction of the rainforest is predictable. Governments hoping to build dams and create hydroelectric facilities; to build highways, railroads, and power lines; and to channel rivers to expand economic development further threaten the survival of the forest. Scientists are concerned that forest loss could also escalate in the Amazon as the climate becomes increasingly dry—in part because of forest clearing.

Without a concerted effort by local communities, governments, and international organizations, only a small fragment of the rainforest will survive. These fragments cannot hold the type of diversity we presently find within the Neotropical rainforest and the loss of diversity seems inevitable.

The Afrotropical Rainforest (Africa and Madagascar)

The African expression of the Tropical Rainforest Biome occurs in two main areas, across the lowlands of Central Africa and along the coastal regions of West Africa. A fragment of rainforest persists in a strip on the eastern coast of Madagascar. The Afrotropical rainforest or African tropical rainforest is concentrated between 8° N and 8° S latitude. The Afrotropical rainforest includes 18 percent of the world's Tropical Rainforest Biome. Currently, about 0.72 million mi^2 (188 million ha) of rainforest are left in Africa, where, especially in West Africa, they have been depleted by commercial logging and conversion to agriculture. In the past, the African tropical rainforests formed a more or less continuous forested area along the west coast of Africa from Sierra Leone through the Democratic Republic of the Congo (the DRC, formerly Zaire) and east into Uganda. Fragments of montane rainforest still exist in western Kenya, Rwanda, and Tanzania. The rainforests in West Africa and Central Africa are separated by the Dahomey Gap, an area of savannas and dry woodlands in Togo, Benin, and eastern Ghana. Rainforests cover the lowlands at elevations below 3,300 ft (1,000 m); most areas are less than 700 ft (200 m) above sea level. Rainforest is present in the countries of Cameroon, Central African Republic, Congo, the DRC, Equatorial Guinea, Gabon, the Gambia,

Ghana, Guinea-Bissau, Côte d'Ivoire, Kenya, Liberia, Nigeria, Rwanda, Senegal, Sierra Leone, Tanzania, Uganda, Zambia, the Seychelles, and the islands of São Tomé and Principe in the Gulf of Guinea (see Figure 3.9). The rainforest of Madagascar is also part of this biogeographical region.

In West Africa, nearly 90 percent of the original rainforest is gone, and the remainder is heavily fragmented and in poor condition. More than 70 percent of Africa's remaining rainforests are located in Central Africa within the remote areas of the Congo Basin in the DRC, Uganda, Republic of Congo, Central African Republic, Equatorial Guinea, Gabon, and eastern Cameroon. The Congo River (also called the Zaire) is the Earth's second largest river by volume, and its basin houses the world's second largest rainforest. The rainforest of the Congo Basin is one of the world's most endangered ecosystems. It holds a large part of Africa's biodiversity. Logging, subsistence and export agriculture, cattle ranching, and widespread civil fighting have destroyed forests and displaced native peoples. The expansion of the bushmeat trade threatens the survival of many species. Since the 1980s, Africa has had the highest deforestation rates of any region on the globe. The Congo

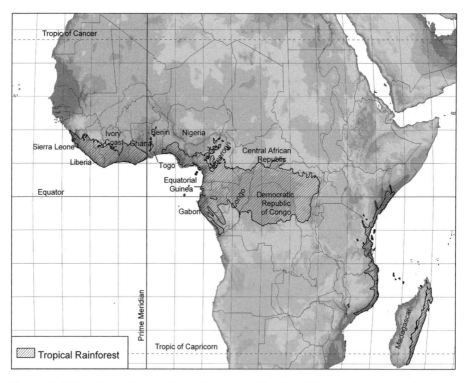

Figure 3.9 Location of tropical rainforests within the African region. *(Map by Bernd Kuennecke.)*

Basin remains the last large remnant of a once continuous and vast expanse of tropical rainforest.

Origins of the African Rainforest

The tropical rainforests of Africa reflect the history of the continent. Tens of millions of years have shaped the African Rainforest. Underlying a large part of the region is one of the oldest rock formations in the world, the 500-million-year-old Precambrian Shield. This ancient basement rock occurs in pieces in Africa, as well as in Asia and South America. Significant tectonic events and severe climatic episodes have influenced the current distribution of the rainforest. Ancient plant families known only from the fossil record in Africa have living relatives in South America, Asia, and Australia, providing evidence of the Gondwanan connections that once existed among these continents.

The modern African rainforest appeared about 35 mya. At that time, rainforests covered a substantial portion of the continent, stretching north to the Arab Republic of Egypt and Libya and east to the Indian Ocean, in areas now covered by desert and savanna. Fossil pollen found in those areas shows the presence of tropical forest species. After separating from South America, Africa was isolated and a unique group of plants and animals evolved. As the African continent continued to move northward, it collided with Asia, about 20 mya. It remained isolated, however, due to climatic barriers imposed by large deserts.

The rainforest reached its largest extent during the Miocene (35–10 mya), and then global cooling brought cooler and drier climates to Africa. This resulted in a decrease in the extent of the African rainforest and an expansion of the savannas and woodlands seen in Africa today.

Another period of global cooling occurred about 2.5 mya and brought a cooler, drier, and more seasonal climate that further restricted the African rainforest. Cycles of glaciations continued throughout the Pleistocene Epoch. The last major glacial expansion ended about 18,000 years ago. Studies of fossil pollen (palynology) have found species more accustomed to drier environments, lowered lake levels, and increased dune activity during these times.

The effects of past global climate change on rainforest distribution in Africa were severe. Researchers estimate that the African rainforest was reduced to as little as 10 percent of its largest, Miocene, extent. It only remained in small isolated blocks with higher rainfall and along river margins. These areas became refugia where once widespread animals or plants were able to survive.

Madagascar was isolated from Africa around 90 mya and developed its own unique tropical rainforest along its eastern coast. Although limited in extent, Madagascar is home to some of the richest rainforests on Earth, and more than half of all Madagascar's rare and wonderful plants and animals are found in these forests.

Climate

The climate of the African rainforest is consistently warm and humid with little seasonal or annual fluctuation. Temperatures average between 70° and 90° F (21° and 32° C) during the day, with cooler temperatures along the coast. Mean monthly minimum temperatures show little variation, ranging from 57°–63° F (14°–17° C), with maximum temperatures reaching 84°–90° F (29°–32° C). Humidity remains at least 70 percent and is often 90–100 percent. The climate inside the rainforest has a smaller temperature range than outside the forest.

Most African rainforests are drier than those in other regions and have annual rainfall totals of 60–80 in (1,500–2,500 mm). Rain occurs most months, with no more than two or three months receiving less than 4 in (100 mm). In certain areas, rainfall can be extreme; for example, 470 in (12,000 mm) was recorded in the forest on Mount Cameroon. Although rainfall is abundant, it is the limiting factor in rainforest distribution. Tropical rainforests give way to tropical seasonal forests, woodlands, and savannas when a distinct dry season brings rainfall of less than 4 in (100 mm) for several months.

Although day length within the African rainforest changes little throughout the year, cloud cover provides a significant variation in the amount of solar radiation reaching the forest. In the rainiest season, there can be as little as two hours of sunlight directly reaching the forest. Even in the drier season, the forest remains misty.

The global circulation that heavily influences the African rainforest is the shifting ITCZ. In Africa, a maritime airmass flowing southwest from the Atlantic Ocean meets an opposing hot, dry continental airmass from the northeastern deserts. Where the two meet is a zone of instability and high rainfall. Airmasses move seasonally from north to south with the shifting ITCZ, varying from 5°–7° N latitude in January to 17°–21° S latitude in July. These movements account for the seasonal distribution of rainfall in tropical Africa. To the south of this zone is an area of continuous low clouds, but less precipitation. The Trade Winds are less of a factor in the African rainforest than in the Neotropics, but they may be important in the dispersal of species. From December to March, hot dry winds known as the Harmattan Winds blow from the Sahara Desert, reaching the rainforests of West Africa. In the Congo Basin, dry winds from the Ethiopian Highlands blow over the rainforest. This is less severe than the winds encountered in West Africa. Both winds bring warm dry air and decreased rainfall to the rainforests.

General forecasts about the future African climate in face of global warming predict increased aridity in equatorial Africa, with climate shifts causing the migration of vegetation zones and limiting the distribution of tropical rainforest. Slight temperature variations can cause significant effects on animal composition, which in turn affect plant pollination and seed dispersal with consequent deleterious effects on the forest.

Soils

Soils tend to be old and nutrient poor, originating from ancient bedrock of the Precambrian shield. More than 55 percent of the soils in the African rainforest are classified as either oxisols or ultisols.

Forty percent of soils found in the Tropical Rainforest Biome in Africa are oxisols. Oxisols in Africa have a characteristic deep yellow or red color due to high iron oxide content caused by heat and heavy moisture. They are concentrated in Gabon, DRC, Rwanda, Burundi, Liberia, Sierra Leone, and eastern Madagascar.

Other common tropical soils in the Afrotropical region are ultisols. They make up about 16 percent of African rainforest soils and are red and yellow in color. They are somewhat well drained and typically found on slopes. In Africa, ultisols occur in the eastern Congo region, on forested zones of Sierra Leone, as well as in parts of Liberia, and along a thin strip from Côte d'Ivoire to Nigeria.

The other soil types are divided among entisols, inceptisols, and alfisols. A type of entisols called psamments covers 15 percent of the African rainforest. These are deep, sandy soils with high acidity and very low fertility, which are susceptible to erosion. The largest extent of these soils occurs in the western portion of the Congo Basin in the DRC and the west Central African Republic. Inceptisols represent 12 percent of tropical soils. They are neutral soils that are less weathered than the oxisols and ultisols and have clearly developed A, B, and C horizons. Since they are relatively fertile, they are used for agriculture.

Vegetation

Plant diversity is high in the tropical rainforests of Africa, but lower than in the other two rainforest regions. This is especially evident in the limited number of palms, lianas, and orchids and other epiphytes. The relative low species diversity is probably due to the extensive dry periods in the geologic past that severely restricted rainforest distribution.

Africa shares some dominant plant families with the Neotropical and Asian-Pacific rainforests, including the legume (Fabaceae), euphorb (Euphorbiaceae), soursap (Annononaceae), laurel (Lauraceae), mahogany (Meliaceae), palm (Arecaceae), and fig (Moraceae) families. The once-connected continents of Gondwana influenced many of these families' evolutionary histories. Others, however, may be modern in origin and represent more recent dispersal events.

The relationship of African plants with the Asian flora is stronger than with the Neotropics. Africa shares 164 of its 177 families with tropical Asia, including Malaysia. In Madagascar, 200 of 234 families are shared with Asia. Although many trees are common to the African and Asian-Pacific rainforests, *Symphonia globulifera*, commonly called manil or cerrillo, is the only one common to the

Neotropical and African rainforests. This tree is widely distributed in the rainforest of the Neotropics, but limited in the rainforest of West Africa. It is pollinated by birds in both regions. Cerrillo has its origin in Africa about 45 mya and is thought to have come to the Americas 15–18 mya.

Forest Structure

The structure of the African rainforest is similar to other tropical forests with three canopy layers, a shrub layer, and a herbaceous layer (see Figure 3.10). Emergent trees in this region are slightly shorter on average than those of the Neotropical and Asian-Pacific rainforests. Emergent trees average 80–150 ft (24–45 m) in height. Unlike the Neotropical rainforest, the emergent layer in the African rainforest is often dominated by only a few species within a forest stand. Emergents tend to be shade tolerant when young and grow slowly under the canopy until a clearing opens, at which point their growth rate increases and they rapidly fill the gap. Common trees in the emergent layer are members of the Caesalpinioideae subfamily of the legume family and include tamarind, mopane, senna, and honey locusts. All are tall and long-lived, some reaching ages of 1,000 years, although older trees are now rare due to selective logging and deforestation. A single species often dominates large areas. Single-species dominance is common in the African rainforest, but it is less common in other regions within the Tropical Rainforest Biome. While an acre of rainforest in the Neotropics may contain hundreds of tree species, that same area in Africa may have less than 20. Trees found in the tropical rainforests

Figure 3.10 The structure of the African rainforest is similar to the other regions, although lianas play a less significant role. *(Illustration by Jeff Dixon. Adapted from Richards 1952.)*

of the Neotropics and Asia have limited distribution, but the trees of the African rainforest occur throughout the region.

The B layer has an open canopy formed by 50–100 ft (15–30 m) trees. Trees in this layer tend to be narrow. Woody vines and epiphytes found among the B-layer trees can make up a majority of the layer's biomass. The C layer is composed of smaller trees around 30–50 ft (5–15 m) tall. The branches of many trees in the B and C layers are often intertwined, providing arboreal avenues for animals traveling from one location to another. Together, the two layers form a closed canopy (see Figure 3.11). The trees of both layers include members of the palm (Arecaceae), laurel (Lauraceae), aroid (Araceae), pepper tree (Piperaceae), and melastome (Melostomataceae) families. The kola nut tree is an interesting lower canopy tree from the tropical rainforest of West Africa. The seeds from this tree have been used for centuries as a medicinal tonic.

African rainforests tend to have lower tree densities (the number of trees per unit area) than the rainforests of the Neotropics or Asian Pacific. Studies reveal tree densities in tropical rainforests range from 300 to 1,000 trees per 2.4 acres (1 ha). African densities tend to be on the lower side of this number, with 300–600 trees per hectare. The canopy of the African rainforest is rich in flowering trees, orchids,

Figure 3.11 African rainforest on São Tomé Island, Republic of São Tomé and Principe. *(Photo courtesy of Robert Drewes, Ph.D., California Academy of Sciences.)*

Kola Nuts

Kola nuts are the seeds of an evergreen tropical rainforest tree (*Sterculia acuminata*) in the family Sterculiaceae, the same family as cacao. Kola nut trees can grow to 60 ft (18 m) tall. They are indigenous to West Africa, growing in the forests of Sierra Leone, Liberia, Côte d'Ivoire, and Nigeria. They also grow along rivers in Central Africa in Gabon and the Congo River Basin.

The fruit of the kola nut tree forms a star-shaped cluster of pods, with each pod enclosing 4–10 chestnut-size seeds. Kola nuts have important uses in traditional medicines and ceremonies and are cultivated for local and export markets. They are used by indigenous people as a stimulant and are high in caffeine (2.5–3 percent) as well as theobroma. Kola nuts are also used as an antidepressant and are thought to reduce hunger and fatigue, aid in digestion, and work as an aphrodisiac. In some West African cultures, kola nuts are given as gifts or expressions of hospitality. They convey a sign of respect or an offering of peace.

Kola nuts were used in the early production of many tonics and soft drinks, such as Coca Cola. Even today, the kola nut is used in the production of natural cola drinks and as an extract used in natural medicines. Kola nuts are produced commercially in Africa and South America (where they were introduced from Africa) for export into Europe and North America for use as a flavoring. However, most of the trade in kola nuts is concentrated within Africa.

vines, and other light-loving plants, but it is not nearly as diverse as the other rainforests of the world.

Since little light penetrates beyond the upper canopy layers, the plants in the shrub layer must be able to survive under limited light conditions. Certain families of shrubs are common in the understory. They include the smaller trees and shrubs of the melastome, coffee, composite, legume, and pepper tree families previously mentioned above, along with nettles (Urticaceae), ferns, and seedlings and saplings of the higher canopy trees.

The forest surface is sparsely populated with ferns, herbs, and seedlings of canopy trees. The ground layer is composed mostly of dead, decaying plants and animals, and the organisms that decompose this material. The layer of decomposing material makes the recycling of nutrients throughout the forest possible. It is a crucial facet of the African rainforest. Ferns, mosses, orchids, ginger (Zingiberaceae), sedges (Cyperaceae), aroids (Aracaceae), and African violets (Gesneriaceae) are common in Africa as well as other rainforests.

African Rainforest Trees

Trees of the African rainforest share characteristics with tropical trees in other tropical regions. These trees tend to have thin bark, usually less than 0.5 in (1.2 cm) thick, in a variety of colors from light to dark, with or without splotches. Tree bark is typically smooth to discourage the growth of vines on the tree. Some bark is thorny to discourage animals. The wood is often hard and dense as protection against the wood-eating insects. This adaptation makes the wood a valuable hardwood resource. Like in other rainforests, many emergent trees found in the African rainforest have buttresses at the base of the trunk. These provide support for tall trees that typically have a shallow network of roots. Buttresses can be 15–32 ft (5–10 m) high where they join the trunk. Aerial or stilt roots are also common, particularly in palm species.

Although flowering can occur throughout the year, fruiting typically occurs over a multiyear cycle. The result is intervals of mass production of several tons of large edible fruits and nutritious seeds, the majority of which are destroyed by invertebrates. These fruits and seeds are crucial to the survival of animals in the forest. In turn, animals are critical for the germination and dispersal of many of these seeds. Although insects pollinate most African rainforest plants, 75 percent of their seeds are dispersed by mammals, birds, and fish.

Many of the seeds of the African rainforest are covered by hard shells and surrounded by fruits enclosed in tough casings. Vertebrates eat many of these fruits. Indeed some seeds need to go through an animal's digestion system to break the seedcoat for germination. The makore tree, for example, is dependent on forest elephants to crush the large, hard seed case and consume the seeds. The elephants efficiently disperse the seeds that easily germinate in their dung.

Some African rainforest trees and lianas exhibit cauliflory, where flowers and fruits grow from the trunk rather than off tree branches. False kola is an African tree displaying cauliflory. Several figs also exhibit cauliflory.

Vines are an important structural feature within the African rainforest and the leaves and fruit are a food source for animals. The African rainforest is relatively poor in lianas compared with its regional counterparts. Several plant families, including the legume and coffee families as well as dogbane (Apocynaceae), squash (Cucurbitaceae), morning glory (Convolvulaceae), milkweed (Asclepiadaceae), knotweed (Polygonaceae), among others have lianas and vine species. Fig trees are common stranglers in the African rainforest as well as other tropical forests of the world, using their host trees for structural support.

Epiphytes are present in the African rainforest, but diversity and abundance is less than in the other rainforest regions. At least twice as many genera and species are found in the Neotropical and Asian-Pacific rainforests. The lower number of epiphytes correlates to the relatively lower rainfall amounts in this region as well as to the limited areas of rainforest. African epiphytes consist primarily of orchids and ferns with a notable absence of the bromeliads so common in the Neotropics. More than 100 epiphytic species have been identified within the African rainforest. Orchids are relatively abundant in the African rainforest, with about 15 percent of the world's orchid species occurring here.

Swamp Forests

The swamp forests of the Niger and Congo Delta are the largest such habitats in Africa. Precipitation is high at 100–160 in (2,500–4,000 mm) on average, with an average annual temperature of 82° F (28° C). Hydrology determines species composition within the swamp forests. Flooded forests occur in lower-salinity areas behind mangroves, along rivers, and in low areas of poor drainage. Species in these areas depend on the depth, frequency, and duration of flooding. Many flooded forests are dominated by one species. Plants of the flooded forest typically have large seeds that float to facilitate dispersal and germination. Fish living within these flooded forests assist in these processes. The threatened Niger Delta pygmy hippo, crested genet, and chimpanzee are found in these forests.

Madagascar

Madagascar has a small rainforest, but one quite high in biodiversity. Located along the east coast of the island country at elevations between 0 and 2,600 ft (0 and 800 m), the forest experiences warm temperatures, high humidity, and more than 80 in (2,000 mm) of rainfall annually.

Like other rainforests, five forest layers appear in the forests of Madagascar. The canopy is rather low (80–100 ft [25–30 m]), with high tree density and a mix of tree species. Unlike Africa, no single tree dominates the canopy. The trees are typically straight, their branches covered with lianas and epiphytes. The canopy is not strictly closed; most trees have spaces between their crowns. Canopy-dwelling animals must be able to negotiate these gaps by climbing, leaping, gliding, or flying.

Dominant trees are members of the legume (Fabaceae), mangosteen (Clusiaceae), ebony (Ebenaceae), palm (Arecaceae), euphorb (Euphorbiacea), sapodilla (Sapotaceae), frankincense (Bursiaceae), and elaeocarpus (Elaeocarpaceae) families. Elaeocarps occur in Madagascar and throughout Asia, but not in Africa. More than 170 palms are present, many related to Asian palms. Pandans (Pandanaceae) and bamboos (Poaceae) are abundant. More than 90 percent of the trees and shrubs in the rainforest are endemic.

Animals of the African Rainforests

The animals of the African rainforest are quite diverse with species found nowhere else in the world. New species continue to be discovered within the African rainforest. As exploration into less-accessible forested areas continues, a fuller understanding of the animal life will emerge. The continuation of taxonomic and evolutionary analyses of newly discovered animals is also important in this understanding.

Mammals

More than 270 species of mammals, belonging to 120 genera, and more than 25 different families have been identified in the lowland rainforests of Africa (see Table 3.3). The African rainforest contains an exceptional diversity and abundance of large mammals, unmatched by any other tropical rainforest. Questions remain regarding the geographic origin and evolutionary relationships of some mammals present in the African rainforest, particularly regarding their relation to South American mammals.

Several reasons exist for the remarkable diversity of animals. Stratified forest layers result in a multitude of habitats in the canopy and on the ground. Forest resources are remarkably varied, providing different opportunities for different mammals. Numerous plants provide leaves, shoots, flowers, fruits, seeds, and bark, as well as insects for consumption. These resources are abundant and available throughout the year. Mammal distributions are closely correlated with the different vegetation layers. The upper canopy houses primates, rodents, bats, pangolins, tree

Table 3.3 Mammals Found in the Rainforests of Africa and Madagascar

Order	Family	Common Names
Afrosoricida	Tenrecidae	Tenrecs and otter shrews
Artiodactyla	Boviidae	Antelope
	Giraffidae	Okapis
	Hippotomidae	Hippos
	Suidae	Pigs
	Tragulidae	Mouse deer, chevrotains
Carnivores	Felidae	Leopards and golden cats
	Herpestidae	Mongoose
	Mustelidae	Otters and ratels
	Viveridae	Civets and genets
Chiroptera	Emballonuridae	Sac-winged, sheath-tailed bats
	Hipposideridae	Leaf-nosed bats
	Megadermatidae	False vampire bats
	Molossidae	Free-tailed bats
	Myzopodidae	Old World sucker-footed bats
	Nycteridae	Slit-faced bats
	Pteropodidae	Old World fruit bats
	Rhinolophidae	Horseshoe bats
	Rhinopomatidae	Mouse-tailed bats
	Vespertilionidae	Insect-eating bats
Hyracoidea	Procaaaviidae	Hyraxes
Insectivora	Scoricidae	Shrews
Macroscelidea	Macroscelidae	Elephant shrews
Pholidota	Manidae	Pangolins
Primates	Cercopithecidae	Old World monkeys
	Cheirogaleidae[a]	Dwarf lemurs
	Daubentoniidae[a]	Aye ayes
	Galagidae	Galagos
	Hominidae	Great apes
	Indriidae[a]	Indris, sifakas
	Lemuridae[a]	True lemurs
	Lepilemuridae[a]	Sportive lemurs
	Lorisidae	Pottos
Proboscidea	Elephantidae	Elephants
Rodents	Anomaluridae	Flying squirrels
	Gliridae	Door mice
	Hystricidae	Old World porcupine
	Muridae	Old World rats and mice
	Nesomyidae	African and Malagasy endemic mice, rats
	Sciuridae	Squirrels

Note: [a]Found only in Madagascar.

hyraxes, and flying squirrels that seldom leave the trees. The middle canopy zone includes arboreal mice, squirrels, bats, genets, and primates. On the ground, an entirely different group of animals is found.

Africa has the largest number of ground-dwelling rainforest mammals, the largest being the forest elephant, buffalo, bongo, okapi, and leopard. Medium-size terrestrial mammals include rodents, mongoose, pangolins, duikers, the smaller cats, and otters. The great apes—gorillas, chimpanzees, and bonobos—spend time on the ground as well as in the trees. All African mammals are placental mammals. There is no evidence that egg-laying monotremes or pouched marsupials were ever in Africa.

Pangolins are solitary, nocturnal mammals that feed mostly on ants and termites. They are found in the tropical regions of Africa and Asia. Four of the seven pangolin species occur only in Africa. They range in size from the smallest, the tree pangolin, weighing about 3.5 lbs (1.6 kg), to the largest, the giant pangolin, weighing up to 72 lbs (33 kg). All have extremely long tongues and no teeth. Pangolins' backs are covered with a series of overlapping scales that extend from the head, down the back, and almost to the tip of their long tail, while their undersides are covered with a sparse coat of fur. They have short limbs adapted for digging and large sharp recurved claws. Two species have semiprehensile tails used in climbing. When threatened, pangolins curl into a ball, protecting their scaleless undersurface. They lash about dangerously with their tails, which are covered with sharp-edged scales. They may also spray a foul-smelling liquid from their anal glands. Pangolins have relatively poor vision and hearing and locate their prey by scent. Two small pangolins, the tree pangolin and long-tailed pangolin, spend most of their time in the trees. The ground pangolin and the giant pangolin (the largest of all pangolins) are primarily ground-dwelling species. Pangolins are hunted for their meat and their populations are in severe decline. As an example of convergent evolution, pangolins are somewhat similar in appearance and behavior to Neotropical armadillos, but they are not related.

Tenrecs are limited in distribution. A few species live in the rainforest of West and Central Africa. The largest and most diverse group occurs on the island of Madagascar. The family has been divided into three subfamilies. The otter shrews (Potamogalinae) live in West and Central Africa; the tenrecs (Tenrecinae) and the aquatic, rice, and shrew tenrecs (Oryzoryctinae) are confined to Madagascar. Tenrecs are considered direct descendants of primitive mammals as they maintain many of those characteristics.

Otter shrews are only found in the tropical rainforests of Africa and all are aquatic. Otter shrews are large, around 24 in (600 mm) in length and 2.2 lbs (1 kg) in weight, and are adapted for aquatic life. Although called a shrew, the giant otter shrew is really a tenrec and probably represents an early branch of the tenrec family. The giant otter shrew is found only in Central Africa and lives in swamps, streams, rivers, and forest pools within the rainforest. The giant otter shrew is nocturnal, hunting between dusk and daybreak. A strong swimmer, it feeds on crabs, frogs, fish, and eats insects, mollusks, and prawns. The other two tenrec subfamilies occur in Madagascar, where tenrecs have adapted to fill many of the niches taken by rodents, shrews, opossums, and even otters on the mainland. They are aquatic, arboreal, terrestrial, or fossorial. Members of the tenrec subfamily can be relatively large (up to the size of a cat) and highly variable. Most have barbed and detachable hairs modified into spines or quills. Tenrecs are mostly nocturnal and omnivorous. The members of the aquatic or rice shrew subfamily are small and are more mole or shrew like. They lack spines and some are highly fossorial.

Another group of endemic African mammals are the elephant shrews. They are not related to shrews and are thought to be distant relatives of tenrecs, hyraxes, elephants, manatees, and aardvarks. Nineteen different types of elephant shrews are found across Africa. The checkered and four-toed elephant shrew genera live in the African rainforest. They range from mouse to squirrel size. Elephant shrews have long snouts and large eyes and ears. They are accomplished leapers with their long hind legs and escape from predators with long jumps. They are insectivorous and forage at night. Elephant shrews include fruits, seeds, and other vegetable material in their diets and often create elaborate trail systems through the leaf litter on the forest floor. True shrews are also found in the African rainforest with 150 species described.

Bats are abundant in the African rainforest. Both major groups of bats, Megachiroptera and Microchiroptera, are present. Megachiroptera are large fruit-eating bats with large eyes. They use vision to navigate through the forest. Microchiroptera are the small bats that use echolocation, a kind of bat sonar, to navigate and to locate food.

Bats are nocturnal, and during the day, they will hang by their hind feet in darkened areas, on small branches and leaves, logs, dark holes in trees, caves within the forest, and banana trees on the forest edge. Different species of bats prefer different shelters at different layers within the forest, from the ground to the higher canopy. Fruit bats tend to roost singularly or in small groups while the insect-eating bats can form large colonies. The formation of large, densely packed colonies helps the smaller bats stay warm. As well as eating fruit, dispersing seeds, and eating insects, several bat species pollinate tropical plants.

Rodents are a large and successful order in Africa, as elsewhere. Squirrels, flying squirrels, dormice, and Old World mice and rats are within this order. Several rodents are important consumers of seedlings, seeds, and insects, and play major roles in forest dynamics and regeneration.

Squirrels are the most abundant arboreal rodent in the rainforest of Africa. About two-thirds of all African squirrels live in the rainforest. They range in size from the pygmy squirrel, with a body length of 2.7 in (70 mm) and a tail length of 2.3 in (60 mm), to the forest giant squirrel with a body length of 10–13 in (250–330 mm). Squirrels have long bushy tails and can rapidly run through the canopy, jumping from tree to tree. They live in all layers of the canopy, although many are restricted to one or two of them. Squirrels occupy niches filled by tropical birds, such as parrots, in other regions.

Flying squirrels belong to a different family and have unique adaptations for their strictly arboreal life. They have large fur-covered flaps of skin between their limbs that enables them to glide from treetop to treetop. Some flying squirrels have been recorded gliding as much as 320 ft (100 m) without losing much height. They are also excellent climbers. Flying squirrels are restricted to rainforests with tall, old trees with holes where they can rest during the day. They bear a close resemblance to the flying squirrels of the Asian-Pacific rainforest.

Dormice are arboreal rodents that sleep during the day and climb through the trees at night in search of seeds, nuts, fruit, and shoots. They are small rodents with dense fur, long bushy tails, and front paws that are hand like. Although more common in the savannas, they are present in some forests.

Fewer mice and rats inhabit the African rainforest than are found in other regions. They are thought to have arrived from Southeast Asia in the Pliocene Epoch. By that time, the smaller primates and squirrels had already filled most niches, so rats and mice in the African rainforest tend to be less specialized than in other rainforests.

The brushtail porcupine is the only porcupine living in the African rainforest and is the largest rodent in the forest, weighing around 2.2–8.8 lbs (1–4 kg) and having a body length between 14.5 and 24 in (365 and 600 mm). They are nocturnal, roaming the forest floor at night. They are often hunted for their meat.

Hyraxes are large rodent-like animals with small ears, short legs, and no tails. They live in the canopy and feed on vegetation. Hyraxes are most closely related to elephants and manatees. Like elephants, their upper incisors form small tusks.

Primates of the African rainforest region range in size from the small bushbabies (galagos) weighing less than 3.5 oz (100 g) to gorillas that can weigh 660 lbs (300 kg) or more. Primates are divided into two suborders: the prosimians and the anthropoids or man-like apes. The prosimians are considered the more primitive primates, yet some, like the lemurs, have evolved specialized characteristics. A typical African rainforest community has several prosimians, such as lorises, bushbabies, and pottos on the mainland, and lemurs in Madagascar. All tend to be nocturnal and most are arboreal. Some rainforest communities also include anthropoids. These are catarrhines, distinct from the platyrrhine primates of the Neotropics. Unlike their Neotropical cousins, catarrhine primates have a reduced number of premolars, downward-facing nostrils, and flattened nails. Most have nonprehensile tails. Catarrhine primates include Old World monkeys (Cercopithecoidea) and great apes (Hominoidea). The Old World monkeys are medium to large in size and include macaques, guenons, vervets, Diana monkeys, colobus monkeys, and baboons. Many spend their lives in the trees and seldom descend to the ground.

Colobus monkeys live high in the canopy and are able to leap great distances. Different from most other Old World monkeys, colobus monkeys do not have opposable thumbs. The black colobus monkey is one of Africa's most endangered species. Deforestation, forest fragmentation, and hunting all contribute to the steadily declining colobus populations.

Three great apes are present in the African Rainforest. The gorilla, chimpanzee, and bonobo differ from monkeys in the length of their limbs, their lack of a tail, and their larger brain cases. Although not as agile as monkeys, they possess greater manipulative and communicative skills. Most feed primarily on plants, but the chimpanzee is known to be carnivorous when living on the savanna. Gorillas are the largest primate in the world. Two species live in the rainforest in different locations. The western gorilla inhabits the lowland rainforests in Gabon, Cameroon, Nigeria, and Congo; and the eastern or mountain gorilla lives in the montane cloud forests of Rwanda, Uganda, and the DRC. Gorillas live in family groups consisting of one dominant silverback male, one to three subadult males, females, and juveniles. Gorillas build nests each night; males usually build theirs on the ground or in low branches, whereas females make nests high up in the trees. Gorillas typically feed on leaves, stems, and bamboo shoots. Despite a reputation for being brutal, gorillas are gentle creatures. Although protected, gorillas are often hunted for meat. During many of the conflicts in Central Africa, rebels and refugees have entered national parks where gorillas live and have killed many of them. Another growing threat to gorillas and chimpanzees is the Ebola virus. Ebola hemorrhagic fever is fatal and has been responsible for the deaths of thousands of gorillas in Central Africa.

Three subfamilies of chimpanzees live in the African rainforest. The eastern chimpanzee is found in Central and Eastern Africa and lives in a variety of habitats from dry savanna to rainforests. The common chimpanzee is found in Central African rainforests and open woodland forests. The western chimpanzee is found in the West African riverine forests, semideciduous forests, and rainforests. Depending on subspecies, male chimpanzees can reach 95–132 lbs (43–60 kg) in weight, whereas females are typically smaller at 73–104 lbs (33–47 kg). The chimpanzee is primarily a frugivore, but it also consumes seeds, nuts, flowers, leaves, pith, honey, insects, eggs, and vertebrates, including monkeys. During the day, the males will travel and hunt in groups; females travel alone or with their offspring eating insects, eggs, and fruits and vegetation along the way. Every night they construct nests of branches and leaves in a tree. Chimpanzees will use an array of tools in hunting, preparing food, and grooming. Chimpanzees also have complex social interactions and elaborate communication skills.

Bonobos are sometimes called pygmy chimpanzees. They are found only in the rainforest of the DRC, south of the Congo River. Their fur is black, but turns gray with age. Bonobos males weigh about 85 lbs (38 kg) on average. Females are smaller, weighing 66 lbs (30 kg). Bonobos tend to be more slender than chimpanzees. The bonobo is mostly a frugivorous species, but when fruits are unavailable, they eat leaves, flowers, seeds, bark, herbs, invertebrates, and small vertebrates. Some of the vertebrates taken are flying squirrels and young forest duikers. Bonobos live primarily in the forest trees in groups of 50 to 100 individuals. Like the other great apes, at night each individual will make a nest. Bonobos are thought to be the primate most closely related to humans. They display intelligence, emotionality, and sensitivity. Bonobos and humans share 98.4 percent of the same genetic information (DNA). Although hunting bonobos is illegal, poaching continues to have a dramatic effect on their populations. Numbers of bonobos have drastically fallen, from an estimated 100,000 in 1984 to about 5,000 today. If drastic measures are not taken, this species will soon disappear.

Many species of lemurs live in Madagascar. Lemurs are found nowhere else. Through adaptive radiation, they have filled niches taken by squirrels, monkeys, and sloths, as well as some birds in other forests. Current thought is that Madagascar's lemurs descended from a single species that arrived on the island 65–60 mya. Although many have gone extinct, five families, 14 genera, and 32 species survive on Madagascar. In the rainforest, small nocturnal nest-building lemurs, such as mouse lemurs; medium-size diurnal leaf- and fruit-eating lemurs, such as the ring-tailed lemur; small, nocturnal, arboreal, leaf-eating sportive lemurs; and the larger insectivorous aye ayes can still be found. Like other primates, lemurs play a crucial role in the dispersal and germination of certain rainforest plants.

The majority of large African herbivores belong to the order Artiodactyla, and many of these even-toed ungulates inhabit the rainforest, including okapis, bongos, antelopes, forest buffalo, hippos, forest hogs, and bush pigs.

The okapi is a relative of the giraffe adapted to life in the forest. Okapis look like a cross among antelopes, giraffes, and zebras. They have dark brown or gray coats that help them blend into the dark forests. The okapi's legs have white stripes patterned like a zebra (see Plate X). Okapis feed on shoots of young shrubs. They are solitary animals standing about 5 ft (1.5 m) at shoulder height, and weighing between 450–650 lbs (200–300 kg). Okapis tend to be secretive and occur only in small parts of the rainforest in the DRC.

Several antelopes inhabit the rainforest, including bongos, royal antelopes, and duikers. The bongo is a large antelope reddish-brown in color with white stripes across the back. Bongos have horns that project backward so as not to get caught in the dense foliage of the forest. They use their horns to uproot young trees to feed on the roots. Like okapis, bongos are solitary creatures that can move agilely through the forest unnoticed.

Duikers are small antelopes. Duikers, which means "diving buck," are named for their ability to dart off through the forest and hide in thick brush. Fifteen species live in the tropical rainforests of Africa. Some are widely distributed through the forest, while others are limited in distribution. The Jentink's duiker and zebra duiker are restricted to the remaining rainforest of Liberia and the western Côte d'Ivoire. Duikers have rounded backs and short necks with the head held close to the body. They do not have horns. Their coats vary in color from brown to yellow-brown to reddish. Many are named for their coloration, for example, the red-, gray-, and yellow-backed duikers. They tend to weigh about 145 lbs (65 kg) and stand no taller than 1.5 ft (45 cm). Duikers prefer to browse on shoots and leaves, but some will also eat termites and small birds.

While the more common Nile hippopotamus makes its home primarily in the water, another hippo lives mainly in the forest. The pygmy hippopotamus is much smaller than the Nile hippo and is rather pig like in appearance. Pygmy hippos weigh between 350 and 550 lbs (158 and 250 kg), stand about 2.5 ft (0.8 m) at the shoulder, and are about 5 ft (1.5 m) from head to tail. The pygmy hippo is endemic to West Africa. They are typically solitary animals, living alone or in pairs. Though less aquatic than the larger hippos, they are nonetheless very good swimmers. Their diet consists of roots, fallen fruit, leaves, aquatic plants, succulents, and grasses. Pygmy hippos are extremely rare because of the large-scale conversion of the West African rainforest to agriculture, as well as hunting for the illegal bush-meat trade. They are critically endangered in several countries.

Another large artiodactyl that inhabits the rainforest is the giant forest hog. Giant forest hogs are mainly diurnal animals preferring forest thickets near water. They can weigh 390–605 lbs (177–275 kg) and are 6 ft (1.8 m) from snout to tail. Forest hogs live in groups of 5–15 individuals, roaming the forest for fruits, berries, leaves, roots, grasses, and other plants as well as eggs and carrion. Giant forest hogs are rare and are hunted for their meat, which is used for subsistence as well as commercial trade. Their ivory tusks are also traded. The smaller bush pig is another member of the pig family present in the rainforest. Bush pigs tend to be nocturnal and omnivorous.

Forest elephants are the largest herbivore in the rainforest. Forest elephants are smaller and stockier than African elephants, and their tusks are thinner and straighter (see Figure 3.12). Forest elephants tend to be browsers, feeding on leaves and fruit, while the savanna elephant feeds more on grasses. They move through the forest alone or in small groups. Forest elephants are distributed throughout the tropical rainforest in Central Africa. Populations are in steep decline due to continued destruction of habitat as well as hunting. Elephants are important dispersers for some large-seeded plants. They also damage or kill plants through trampling and uproot or debark trees and branches when feeding. Through these activities, they create gaps in the forest canopy that allow light to penetrate to ground level and stimulate plant growth.

Forest elephants face many threats and need closed forests to survive. When elephants venture beyond the forest to find food, human-elephant conflicts become more common. These typically lead to the killing of the trespassing elephant. Elephants are illegally killed for the international ivory trade. Species protection and conservation of habitat is essential.

Carnivores in the African rainforest include otters and ratels, mongooses, civets and genets, leopards, and smaller cats. The ratel or honey badger is a medium-size nocturnal predator that is skunk like in appearance with a large white stripe across its back. Ratels live in burrows. Their diet consists of small vertebrates as well as bulbs and shoots.

The mongoose family consists of genets, civets, and mongooses. All are medium-size terrestrial carnivores that eat insects and small vertebrates although some include fruit in their diet. Genets are cat-like nocturnal mammals that hunt both on the ground and in trees. Civets are more fox like in appearance and strictly

Figure 3.12 Forest elephants tend to be smaller than the African elephant of the savannas. Taken in the Dzanga-Ndoki National Park in Central African Republic. *(Photo courtesy of Brian L. Fisher, California Academy of Sciences.)*

terrestrial. The rare water civet is found only within the DRC. It feeds primarily on fish. The palm civet is more cat like in appearance than the other civets and is mainly arboreal, hunting at night for invertebrates and small vertebrates. Mongooses are small brown or gray carnivores with small heads, pointed snouts, and short, rounded ears. Mongooses feed on a wide range of animals, including small mammals, birds, reptiles, eggs, crabs, and a variety of insects. Some species also include tubers, fruits, and berries in their diet. The Gambian mongoose is an endangered species restricted to southern Senegal and Nigeria.

The fossa, a unique member of the mongoose family, only occurs in Madagascar. It is the largest carnivore on the island and considered a "living fossil" because its continued survival is thought possible only because of the island's isolation and lack of competing carnivores. The fossa looks somewhat like a cat with a slender body, short legs, long tail, and reddish-brown fur. It lives in rainforests as well as dry forests and hunts on the ground and in the trees, taking birds, eggs, lemurs, rodents, and invertebrates.

Several medium to large cats live in the rainforest. The African golden cat is found throughout the equatorial rainforests of Africa. About twice the size of a domestic cat, the golden cat is rarely seen in the wild. They are primarily nocturnal

••

Forest Elephants and Plant Survival

Controversy remains as to whether the forest elephant is a subspecies or an entirely different species from the African elephant of the savanna. Recent DNA evidence suggests that the elephants
are genetically different enough to warrant species designation, yet the debate continues.
Although once abundant, populations have severely declined. Only very small, probably unsustainable groups exist in West Africa, with only 7 percent of its former range intact. Larger populations remain in the Congo Basin of Central Africa.

Forest elephants are opportunistic feeders. Some favor particular fruits and journey distances
to find them. Often, many elephant trails lead to the same majestic makore tree (*Tieghemella heck-
elii*), where elephants gather until fruit is no longer available. The makore tree produces large,
hard-shelled fruit with thick pits that may depend on elephants for germination. In one area of
Ghana, where forest elephants were eliminated, makore seedlings are no longer found. Another
fruit favored by elephants is that of the strychnine plant, a liana that grows high in the canopy. Its
fruit is a green-yellow sphere about the size of a pumpkin and so hard that few animals can consume it. Elephants are able to crack the shell and expose the seeds. The elephant consumes the
seeds, which have an intoxicating effect on them. Defecated seeds germinate in a fertilizer of elephant dung.

Plants dependent on elephants for dispersal and germination have nutrient-rich fruits with
strong odors and hard shells that attract elephants. Elephants help with seed dispersal and germination in up to 30 percent of West African trees. The Guinea plum tree, wild mango, and the panda
plant, among others, have large pitted fruits dispersed by elephants that germinate in elephant
dung. With decreasing numbers of elephants, the loss of many forest trees seems inevitable.

••

hunters of rodents, tree hyraxes, birds, and duikers and rest in the trees during the
day. The leopard is the largest carnivore in the tropical rainforests of Africa. Leopards are solitary, nocturnal predators. Their fur can range from straw-colored with
black spots to almost completely black. The spotted patterns act as camouflage in
the sun-dappled forest. Leopards stalk their prey and then quickly rush to catch it.
They can run up to 60 mph (96 kph) in short bursts. When the hunt is successful,
the leopard will drag its prey into the bushes or up onto a tree branch to keep it
away from other predators and scavengers.

Birds

A variety of birds live within the African rainforest, yet fewer birds inhabit a comparable area of the African rainforest than in the Neotropics, Asia, or New Guinea.
The dominant families include cuckoos, kingfishers, hornbills, bulbuls, shrikes,
Old World Warblers, flycatchers, sunbirds, and weavers. None of these birds occur
in the Neotropical rainforest, but most are represented in the Asian-Pacific forests.

The birds of tropical Africa that eat fruits, nuts, and seeds tend to be large and
brightly colored. Large size lets the bird consume larger fruits and more fruits in

any given feeding. Large size is also effective in reducing the risk of predation by limiting the types of predators that can attack an individual bird. Lowered predation risk has allowed fruit-eating birds to evolve striking color patterns. These colors probably are used to attract mates, as well as to signal potential rivals. They may also camouflage the bird amid the brightly colored flowers and fruits in the canopy.

Turacos are the only bird family endemic to Africa. All 23 species are frugivores, specializing particularly on figs. They also consume leaves, buds, and flowers. They usually feed in mixed flocks with other birds such as green pigeons, hornbills, and barbets. Turacos have small rounded wings and long tails, which make them weak fliers; but their strong legs allow them to move freely within the forest trees. Most turacos are brightly colored with crested heads (see Plate XI). The largest species is the Great Blue Turaco, which measures 30 in (750 mm) in length.

Another brightly colored frugivore in the African rainforest is the hornbill. The hornbill bears a close resemblance to the toucan of the Neotropics, but the two are not related. Hornbills have large heavy bills with bright coloration used to reach fruit at some distance from their perch. The bills are also used in display, to build nests, and for defense. Besides fruit, some hornbills eat insects and other small animals.

The African rainforest has few parrot species, probably due to the large number of squirrels and other rodents that take up typical parrot roles as fruit and seed eaters. The African Gray Parrot is a popular bird in the international pet trade because of its incredible vocal abilities and the speed with which it learns to mimic sounds. Few parrots occur in Madagascar as well.

The other brilliantly colored birds of the African rainforest are nectar feeders. Year-round availability of flowers and nectar provides them with a constant food supply. Sunbirds are the main flower-visiting birds in the African rainforest. They are small and brightly colored, with feathers that can sometimes have a metallic sheen. Many have downward curved beaks and a long tubular tongue that is split in half at the tip. Sunbirds have strong legs and sharp claws to aid in climbing rainforest trees. They usually take nectar while perched, and many plants pollinated by sunbirds have adapted to accommodate their pollinators by producing flowers on their branches in easy reach of perching birds. Africa's great variety of sunbirds occupies similar niches and is similar in appearance to the hummingbirds of the Neotropics, but they are not related.

Insectivorous birds in the tropical rainforest dominate in terms of numbers of both species and individuals. The majority are the small brown birds that forage through the lower canopy and understory, searching for insects and small arthropods on twigs, branches, trunks, vines, and leaves. They often travel in mixed flocks that may include Old World Warblers, flycatchers, babblers, drongos, white eyes, and woodpeckers. Frequently, sunbirds and weavers join the group. Some insectivorous birds gather around ant swarms like antbirds of the Neotropics.

Far fewer ground-dwelling birds live in the rainforests of Africa than in both the Neotropical and Asian-Pacific regions. No large ones inhabit the region. The very large Elephant Bird once inhabited Madagascar, but it is now extinct. Smaller ground birds include Guinea Fowl, Francolins, and the Congo Pea Fowl.

Predatory and scavenger birds have probably been responsible for the antipredator adaptations, such as large group behavior, seen in other birds and in primates. The Crown Eagle is a major predator of monkeys and hunts a range of other mammals and birds. In Madagascar, the Henst Goshawk preys on small lemurs. Other raptors specialize in snakes, lizards, and even wasps. A few owls hunt at night. Shelley's Eagle Owl, the largest of the African owls, captures large nocturnal flying squirrels as its primary prey. Although New World vultures play important roles in American rainforests, no Old World vultures occur in the African rainforest, even though they are abundant in woodlands and savannas.

Birds, along with other animals of the African rainforest, are subject to the impacts of deforestation, land conversion, and increasing human populations. Healthy populations of many of these tropical birds rely on intact rainforests that are quickly disappearing. The commercial bird trade is also responsible for the loss of African birds, particularly parrots.

Reptiles and Amphibians

The warm climate of the African tropics makes it an ideal home for cold-blooded reptiles and amphibians. A great variety of lizards, crocodiles, turtles, chameleons, and snakes, as well frogs, toads, and a few caecilians dwell in the rainforests. Preliminary checklists, brief species descriptions, and notes on natural history provide some clues into the diversity of the herpetofauna, yet little is really known about these species, their evolutionary relationships, or their biogeography. Intensive research in the forest is often difficult because of limited accessibility.

More than a hundred different snakes live in the tropical rainforests of Africa. These rainforests hold the largest, the smallest, and reportedly the fastest snakes in the world. The rock python is reported to be the largest snake in the world, even larger than the anacondas of South America. One rock python found in the rainforest in Côte d'Ivoire measured 33 ft (10 m) in length. Pythons are constricting snakes, literally squeezing the life out of their prey. Other pythons, including the burrowing python and royal python, are found in the rainforest.

Some of the smallest snakes in the world are the worm or blind snakes of Africa. These snakes grow only 6 in (15 cm) in length and have a diameter similar to earthworms. They eat invertebrates and live in termite nests.

Many dangerous snakes inhabit the African rainforest, but most are usually difficult to find. Poisonous snakes belong to the viper, colubrid, or elapid families. The many vipers include golden vipers, rhinoceros vipers, bush vipers, mole vipers, night adders, puff adders, snake eaters, quill-snouted snakes, and tree snakes. The boomslang is a venomous colubrid snake that lives in the trees of the African rainforest. It is the most venomous of the colubrids. Of the elapids, the forest is home to water cobras, forest cobras, and the black spitting cobra, which can atomize its highly poisonous venom and shoot it a distance of 20 ft (6 m).

Other elapids include the green and black mambas, the most dangerous snakes of the African Rainforest. The black mamba is the largest venomous snake in Africa. Its extremely potent venom attacks the nervous system and is 100 percent fatal without antivenom. Black mambas are the fastest snakes in the world and are able to travel 10–12 mph (16–19 kph) in short bursts. Black mambas live in hollow insect mounds, abandoned burrows, and rock crevices. During the day, they actively hunt small mammals, birds, and lizards. Unlike most other snakes, mambas will strike repeatedly if cornered; they have reportedly brought down giraffes and lions with their venom.

The African rainforest is home to a few tortoises and aquatic turtles, including the serrated hinged-back and Home's hinged-backs tortoise, soft-shelled turtles, and side-necked turtles. The number of turtles found in the African rainforest is one-fourth of that found in the Neotropics.

Lizards are perhaps the most common reptiles in the African rainforest. Monitor lizards, agamas, skinks, legless lizards, chameleons, and geckos live in the rainforest. Their food is chiefly invertebrates, but some will eat soft leaves from young plants. Some lizards have developed the ability to drop their tail when caught by an attacking predator; this allows the lizard to escape while its predator is left with the small prize of part of the lizard's tail. The lizard later regrows its tail.

Monitor lizards are the largest lizard in the rainforest, with some 20 species occurring in Africa. The Nile monitor lizard can grow to 6.5 ft (2 m), eating crocodile babies and eggs. Agamas, also known as Old World iguanas, are widespread throughout Africa and Asia. The tropical rainforest is home to several genera that have adapted to life in the warm, wet forest. Agamas are diurnal, relatively large with scaly, spiky bodies and large heads. Most male agamas have brightly colored heads. Skinks all have shiny bodies and long tails. They are fast moving and live on the forest floor. Four legless lizards also occur in tropical Africa.

Geckos are yet another type of lizard. They are small, mostly nocturnal, with soft pale yellow and almost transparent skin. Geckos have toepads with minute hooks that allow them to cling to irregular surfaces. Unlike most lizards, geckos emit sounds such as chirps and barks. Their main diet is insects.

Chameleons differ from lizards in shape and size. They are slow-moving reptiles that live in trees and bushes. Their heads are large with rotating eyes that can move in all directions independently of each other. Along with their prehensile tail, their feet wrap around branches, allowing them a firm grip. Chameleons are best known for the ability to blend into their environment. They can change color, turning from green to brown or brown to yellow, to blend in with the surroundings. Over half of all chameleons are found in Madagascar.

Every part of the African rainforest has its unique community of frogs. Tree frogs are abundant in all rainforests, where they spend most of their time in the trees and shrubs. They differ widely in color, from brown and green, to bright colors with black striping. Tree frogs have sticky discs on their toes, adaptations to life in the trees. They lay their eggs in foam nests between branches or on tall reeds (see

··

Deadly Reptiles

Another large and dangerous reptile of the African Rainforest is the crocodile. Three species of crocodiles live in the wettest forests, in palm swamps, and along open water. The West African dwarf crocodile is the smallest of the three species. It eats small vertebrates, large invertebrates, and crustaceans. The slender-snouted crocodile lives in the wet areas of the tropical rainforests in Central and West Africa. They are medium size and eat mainly fish, amphibians, and crustaceans. The Nile crocodile is the largest African crocodile, reaching lengths of up to 16–20 ft (5–6 m). Adult males weigh about 1,100 lbs (500 kg), with some reaching weights of 2,000 lbs (900 kg). They mainly eat fish, but can also take amphibians, reptiles, birds, and any other vertebrate that comes to the edge of the water. Adult Nile crocodiles have eaten antelope, buffalo, bush pigs, monkeys, cats, other crocodiles, and the occasional human.

··

Figure 3.13). Tree frogs in Africa fill the same niche as the poison dart frogs of the Neotropics, but they are unrelated. Some even produce a neurotoxin on their skin. Reed frogs, rocket frogs, aquatic clawed frogs, and bullfrogs are other frogs found in the rainforest.

Figure 3.13 Rainforest frogs like the oceanic tree frog lay eggs on leaves in the forest where they develop out of the reach of most prey. Taken on São Tomé Island. *(Photo courtesy of Robert Drewes, Ph.D., California Academy of Sciences.)*

Many toads occur in Africa's rainforests. They vary in color; many blend in well with the leaf litter or trees, making them almost invisible, while others can be bright blue, green, or yellow. Toads range in size from tiny to large. Like frogs, many toads are endemic and have limited distributions in forested areas. True toads or Bufonids have been successful in both Africa and South America; Africa is thought to be the origin of these toads with later dispersal into the Americas.

Caecilians are a small group of limbless amphibians that also live in the rainforests of Africa. They resemble worms or small snakes, with shiny skin. Caecilians live in tropical rainforests across South America, Africa, and Asia, suggesting an ancient distribution prior to the breakup of Gondwana. Caecilians typically live below ground or in forest soils or leaf litter and feed on invertebrates.

Insects and Other Invertebrates
The majority of species in the world are invertebrates, yet they remain the least studied. Insects as well as arachnids and crustaceans inhabit the rainforest. While over a million species have been discovered, there are probably millions more to be identified.

Beetles (order Coleoptera) are by far the most diverse order of insects in the African rainforest. While new species continue to be discovered, tens of thousands have yet to be described. Beetles range from 0.25–5 in (1–130 mm) in length. Many have specialized niches. Many eat plants; others are associated with every kind of decomposing matter, while others are parasitic. Some beetles convert wood into dust, clean bones, spread pollen and seeds, enrich the soil, trim leaves and branches, as well as provide a rich protein source for other animals. Scarab beetles, including dung beetles, scavenge on waste and decaying matter. The Goliath beetle is the largest of the African beetles.

Butterflies and moths (order Lepidoptera) are one group that has been well studied in the tropical rainforests of the world. The African rainforest is estimated to house only half the number of butterflies found in the Neotropical or Asian-Pacific rainforests, yet this diversity far outweighs that of all other terrestrial biomes. In the African rainforest, 2,720 butterfly species have been identified. Moths have been studied less than butterflies, so less is known about their species numbers, and life histories. Together, more than 20,000 butterfly and moth species are estimated to occur in the rainforests of Africa. Some butterflies take in nectar while others choose fruit, dung, dead animals, and even animal perspiration as sources of nourishment. The butterfly families that occur in abundance in the African rainforest include the swallowtails, monarch butterflies, the brown butterflies, the whites, the blues, snout butterflies, the nymphs, and the costers that are often brightly colored, unpalatable, or poisonous. None of these families are unique to Africa. Butterflies in the swallowtail family are the largest of the African butterflies. The African giant swallowtail has a wingspan of 10 in (25 cm), only slightly larger than the giant blue swallowtail.

Termites (order Isoptera) are abundant in species, as well as total biomass in the African rainforest. These social insects play a crucial role in maintaining tropical rainforest ecosystems. Tropical Africa has the richest diversity of termites in the world, particularly soil-feeding termites. Their ability to feed on dead plant material makes them vital players in the rainforest. Termites are the dominant decomposer in the lowland rainforest, where they are estimated to consume up to one-third of the annual litter by decomposing it completely or making it more available for other decomposers. Termites themselves are an important food source for some specialized forest species, such as pangolins. Most live underground or in dead wood, but a few species build large mounds or elaborate arboreal nests. Like ants, termites maintain large colonies of a few hundred to several million individuals.

Three main families of termites inhabit the African rainforest: the dry wood termites, damp wood termites, and the so-called higher termites. Dry wood termites, although distributed around the world, seem to be rare in the tropical rainforest. They are thought to have been outcompeted by the higher termite groups. Damp wood termites feed on moist and partially decomposed wood in standing and fallen trees. These lower termites have mutualistic protozoa living in their gut; without them, they could not digest cellulose, a major portion of their diets. The third group of termites, the higher termites, includes 73 percent of all termites in Africa. These termites do not have the protozoa that damp wood termites have, but instead use anaerobic bacteria cultures in their hindgut to assist in food digestion and assimilation. One subfamily in this group is the soil-eating termites. They make up a large portion of the African rainforest termite fauna and are assumed to have an African origin. Many in this subfamily are found nowhere else. Also among the higher termites is the subfamily of fungus-growing termites. These termites consume dead wood and leaves. They have developed a mutualistic relationship with a fungus that grows on termite feces. The fungus breaks down the feces, turning it into a useable food source for the termites. A majority of this subfamily is endemic to Africa. The third subfamily (Nasutitermitinae) is the largest and most specialized group and includes both wood-eating and soil-eating termites. Various species feed on rotten wood, dry or rotten leaves, lichens, mosses, and decomposed organic matter in the soil. Although a large group, they are less abundant and less diverse in Africa than in Neotropical and Asian rainforests.

Ants, bees, and wasps (order Hymenoptera) are also abundant in the tropical rainforests of Africa. They can be found at all layers of the forest and use many different food resources. Army ants present in the Neotropical and African rainforests evolved prior to the breakup of Gondwana. The army ants of both regions share a number of important characteristics. Army ants forage collectively; this increases the types of potential prey available to them, including large invertebrates and small vertebrates. Colonies tend to be nomadic, moving to a new nest when they have exhausted the food supply of the area.

Driver ants are found throughout the forests of West Africa and the Congo. Their role is similar to the army ants of the Neotropics. They have the largest colonies of any social insect. Colonies can consist of tens of millions of worker ants that form long columns with a combined front, traveling at a rate of 65 ft (20 m) per hour, as they swarm through the rainforest in search of prey. They attack other insects, as well as small vertebrates, including birds, and virtually any animal in their path. When a prey item is too large for one ant to carry, the group divides the prey into ant-size pieces and the group returns to the nest with each part. Hunters report that during the rainy season, pythons will search the area for driver ants before consuming prey. Once the snake has eaten a larger animal, it is unable to move quickly, making it more vulnerable to swarms of driver ants.

Weaver ants are canopy ants, spending their entire life within the forest canopy. These ants weave their nests using a plant's leaves and the silk produced by their larvae. Other ant species have evolved particular symbiotic relationships with certain plants in the forest. Some rainforest plants have nectaries outside the flower to facilitate visits by ants. The ants use the nectar, high in carbohydrates, as their food, and the plant as their shelter, and in the process defend the plant against predation by other herbivorous insects. These nectaries are usually located near the most vulnerable part of the plant, such as new leaves and stems.

Bees are important pollinators in the African rainforest. Honeybees are among the most evolved members of the order and are thought to have originated in Africa, where their diversity is greatest. Most honeybees are excellent fliers and if threatened can render a poisonous sting. The Africanized honeybees in other parts of the world are introduced bees or hybrids of African honeybees that are aggressive and quite troublesome.

Arachnids are numerous in the rainforest. Burrowing scorpions, crawling scorpions, flat scorpions, thick-tailed scorpions, whip-tailed scorpions, as well as others live within the rainforest. West Africa is home to the largest scorpion; large and black, the emperor scorpion is 6–8 in (15–20 cm) in length. Most are nocturnal predators. Spiders are also abundant in the rainforest. Spiders are carnivorous, feeding mainly on insects.

Numerous other arthropods play important roles in the life of the tropical rainforests of Africa. Millipedes are scavengers, decomposers of vegetation and animals. Centipedes are carnivorous, feeding on other arthropods, worms, and small vertebrates. Insects such as grasshoppers, crickets, stick and leaf insects, cockroaches, mantids, flies, and fleas all exploit different niches within the forest.

Human Impact on the African Rainforest

Humans have greatly affected the tropical rainforests of Africa. Although humans may have originated in or near the forest, they quickly moved to environments that are more hospitable. It was not until 900 B.C. before they returned to the forest. Most were small groups of people living off the forest by hunting and gathering, or small-scale shifting agricultural groups that seem to have had little impact on the rainforest. Even as recently as 800 years ago, human impact on the forest was minimal. Nontimber uses of the forest included the collection of leaves, fruits, fungi, honey, dyes, resins, gums, and medicines. The hunting and gathering of animals of the forest provided a protein source. Small mammals, snails, and caterpillars were all taken with little to no impact.

As human populations increased, hunting for forest animals increased, but it was not until the twentieth century that larger impacts began to occur. Increased forest settlement, clearing the forest for agricultural exports, and timber production quickly began to limit the forest. Human numbers continued to grow, and further encroachment into the forest took its toll on native plants and animals. Plantations to produce oil palms, rubber trees, cacao, and coffee have replaced the rainforest. Other areas have been cleared for their timber or burned for short-term agriculture.

Most of the tropical rainforests of West Africa have been lost; what remains is late secondary forest. This is especially true in Côte d'Ivoire and Sierra Leone. Liberia retains larger forest blocks, and a few national parks protect some areas. Many forests are held as forest reserves where logging can occur. In West Africa, increased numbers of people are migrating from nearby regions where excessive use has turned the land to desert. Not understanding the fragility of the forest, these immigrants destroy many of the forested areas. Along with the plants, the animals are endangered. Decreased forested land leads to changes in climate, causing further problems at rainforest margins.

Mining for gold, diamonds, and iron ore is also having a major impact on the forests, particularly in the Congo. Oil exploration and drilling are an increased threat. The oil companies' unsafe and environmentally degrading processes have polluted large portions of the forest, especially on the coast and in waterways.

Congo Basin forests are quickly becoming threatened ecosystems. Much of the rainforest destruction in the Congo is due to conversion for subsistence agriculture and the collection of fuelwood by poor farmers and villagers. Access to forested lands follows logging when roads are constructed into the forest. Commercial logging, conversion of forests for agriculture, and widespread civil wars have devastated forests. Illegal hunting continues to increase, causing a sharp decline in many animal populations. As fighting and civil unrest ceases throughout West and Central Africa, governments become more stable, and more areas are opened for logging. Most of the forests not under protection have been allocated to logging concessions. Concessions have even been illegally given in some protected areas. Illegal logging is a significant problem, with corrupt bureaucrats opening restricted areas to cutting in exchange for monetary payments. The sustainability of current logging practices remains questionable. Although the logging industry is a major source of employment in Africa and thousands of workers rely on timber companies for basic health care and other services, cutting timber at the current rate cannot be expected to continue in the long term.

Large-scale human migrations due to armed conflict, tribal warring, and revolutions have brought hundred of thousands of people to the forest seeking protection or peace. However, these people, often out of necessity, are damaging the forest by hunting and clearing land. In some national parks and reserves, staff were threatened or killed, and endangered animals were hunted illegally by warring rebel groups.

Trade in bushmeat is becoming a larger threat, as exploited agricultural lands are no longer able to produce an economical source of protein. Meat from rainforest animals is the primary source of protein for people in villages as well as some cities. The ability of hunters to access the forest is improved by logging roads. Bushmeat hunting is expected to increase as commercial logging expands in the Congo Basin.

As noted earlier, the Ebola virus is killing humans and thousands of gorillas in Central Africa. At least 5,500 gorillas have died from Ebola. This is a greater threat than illegal hunting. Gorillas suffer a 95 percent mortality rate, and chimpanzees have a 77 percent mortality rate. Researchers estimate that Ebola outbreaks over the past 12 years may have killed 25 percent of the world's gorilla population. Efforts to vaccinate gorilla populations are contemplated, but the cost is seen to be prohibitive.

Some hope remains for African rainforests as more countries realize the treasure in their homeland and take steps to conserve the forest and protect its biodiversity. Gabon is working to establish large conservation areas to protect 10 percent of the country. The Republic of Congo created two new protected areas spanning nearly 3,800 mi^2 (1 million ha). Other countries have set aside remaining intact rainforest as national parks and reserves.

Many international environmental organizations abroad, as well as those in Africa, are working diligently to assist in the protection of Africa's last remaining rainforests. In 2007, the World Bank launched a pilot project to avoid deforestation by paying tropical countries to preserve their forests. The $250 million fund will reward countries with tropical rainforest for maintaining their forest to offset greenhouse gas emissions. Tropical deforestation accounts for roughly 20 percent of global greenhouse gas emissions, while an intact rain forest actually absorbs and stores carbon dioxide. Slowing deforestation may be a cost-effective way to slow climate change as well as preserve rainforests and biodiversity.

The African rainforest is essential in maintaining worldwide biodiversity. Although it has been the subject of much research for many decades, there is still more to learn about the forest and its inhabitants. New discoveries are made and new species are identified with each new in-depth study. Stable governments and efforts to promote sustainable development, sustainable forestry, and ecotourism—and to revalue the ecosystem services provided by the forest to Africa and the world—will support current and future conservation in the tropical rainforests of Africa.

The Asian-Pacific Rainforest

The Asian-Pacific expression of the Tropical Rainforest Biome includes 25 percent of the world's total rainforests. The region is also called the Indo-Malayan or

Indo-Asian Rainforest, and the Malesian Floristic Zone. The northern part of the rainforest is located in tropical latitudes, along the west coast of India and Sri Lanka, on continental Asia from Bangladesh and Myanmar (Burma) through southern Thailand, the Laos People's Democratic Republic, Cambodia, and Vietnam, into southeast China and the Philippines. The heartland of the rainforest lies on Malay Peninsula and the islands in the South China Sea, Malaysia, Brunei, Sarawak, and the islands of Indonesia. Rainforest continues on the islands of the Pacific in New Guinea and northeastern Australia and the islands of Melanesia, Micronesia, and Polynesia. The majority of the rainforest lies between 11° N and 11° S latitude (see Figure 3.14). Monsoonal forests and savanna woodlands border the rainforests within this region. In southern China, a belt of tropical rainforest extends to 26° N, nourished by the warm wet winds from the Pacific Ocean. The rainforest in India occurs in two distinct areas, in the northern state of Assam close to the Myanmar border, and in a narrow strip in the Western Ghats along the hills of the west coast on the country. Historically much larger in extent, an estimated 0.61 million mi^2 (156 million ha) of these tropical forests remain.

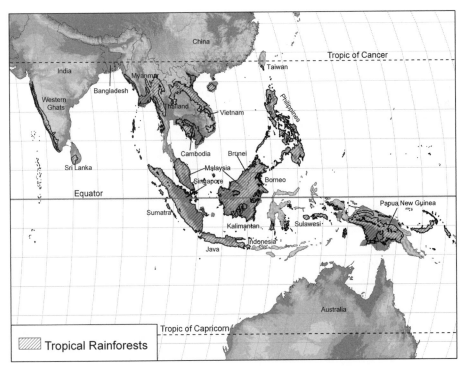

Figure 3.14 Location of tropical rainforests within the Asian-Pacific region. *(Map by Bernd Kuennecke.)*

Origins of the Asian-Pacific Rainforest

As noted in the Chapter 2, during the Tertiary Period (55 mya) the tropical rainforest formed an almost continuous belt from Africa, across Europe into Asia and Southeast Asia, and into the Far East. Although occasionally broken up by narrow seaways, the regions of Africa and Asia were able to share many taxa. By about 10 mya, climatic cooling and drying separated the two areas and led to the development of savannas and deserts, and the connection was lost. During the Pleistocene Epoch, additional cycles of drying and cooling, which significantly restricted the tropical rainforests of Africa and the Neotropics, are thought to have had a lesser effect on the Asian-Pacific rainforest.

More than 40 mya, the island continent of Australia and New Guinea snapped free of the large landmass that included Antarctica and South America and began to drift toward the Equator. Cut off from the rest of the world, the super-island evolved its own distinctive flora and fauna. When the split occurred, the Australian

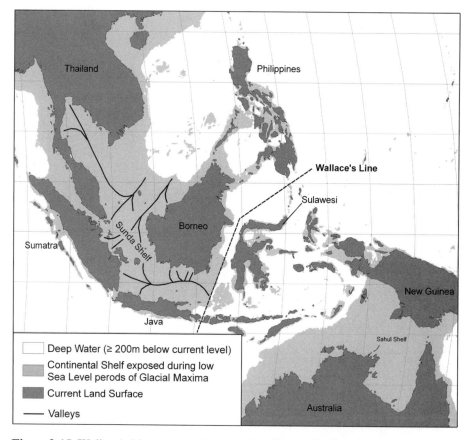

Figure 3.15 Wallace's Line denotes the end of the Sunda Shelf and separates the West and East Malesian regions. *(Map by Bernd Kuennecke.)*

area did not have the moist tropical forest it has today. As the Australian Plate moved equatorward, climate changed. It became warmer and wetter, and the area developed unique plant and animal communities, the precursors of what is found there today. As the Australian region continued its northward movement, the plate collided with the Asian continent (30–20 mya). This allowed for the dispersal of Asian species into the region. Some differences in species between Asian and Australian flora and fauna still exist. Wallace's Line between Borneo (Kalimantan) and Sulawesi (Celebes) and Bali and Lombok farther south define the end of the Sunda Shelf, a shallow extension of the Asian continental plate underlying most of Indonesia (see Figure 3.15). The deep trench at the end of the shelf is a significant barrier to species dispersal, keeping species (particularly mammals) distinctive and separate.

Climate

The climate throughout the Asian-Pacific rainforests is tropical. All seasons are hot and humid, and temperature varies little. Monsoonal winds control much of the climate in this region and create four seasons. A monsoon is a prevailing wind that lasts for several months and then reverses its direction. The southwest monsoonal season occurs from May to September and brings heavy rainfall and strong winds to the rainforests of India and Bangladesh. The southwest monsoon season brings less severe weather to southern Asia and Australasia. From November to March comes the winter, northeast monsoon season. Strong north and northeast winds originating from cold, northern Asia mix with tropical winds and bring severe weather, including heavy rainfall and typhoons, to Southeast Asia and Australasia. These winds are lighter along the Indian continent and produce less rain there. The two other seasons are intermonsoon seasons. During these times, the climate remains hot and humid, but winds and rainfall, although continuous, are lighter than in the monsoonal seasons.

Annual rainfall in the Asian-Pacific rainforest is more than 80 in (2,000 mm). It is distributed throughout the year. In many areas, rainfall can exceed 120 in (3,000 mm), particularly in northwestern Borneo and New Guinea. As noted above, rainfall decreases during intermonsoon seasons, which can last one to four months in some areas. In areas such as New Guinea, no regular dry season develops and monsoonal winds are less significant.

Temperatures average around 87° F (31° C) during the day, with nighttime lows around 72° F (22° C). Average monthly temperatures vary by less than 4° F (2° C), but, as in other rainforests, a single day's temperature can vary as much as 14° F (8° C). Average humidity is 70–80 percent throughout the year.

Changes in the monsoonal cycle can lead to prolonged drought, and periods of hot, dry weather can create an environment vulnerable to fire. In the past, fires have been widespread and catastrophic to the rainforest. El Niño events, which

Wallace's Line

Alfred Russell Wallace was a preeminent bio-geographer of his time. He is best known as the British Victorian naturalist who happened upon the theory of evolution at the same time as Charles Darwin. While Darwin proposed that species evolved because of direct competition with each other, leaving only the fittest to survive, Wallace suggested that the mechanism for evolution was the environment. Wallace traveled extensively throughout the tropics, spending a great deal of time in Southeast Asia and the Pacific. During his travels, he noticed that birds and mammals on the islands of New Guinea and Lombok looked very different from those on the nearby islands of Bali and Borneo to the west. In fact, he noted that animals on New Guinea and Lombok were similar to Australian species, while those on Bali and Borneo were more like Asian species.

Without knowing it, Wallace was seeing the results of tectonic plate activity. New Guinea and Lombok, Timor, Flores, and Sulawesi (Celebes) were part of the Australian plate, whereas the islands to the west were part of the Asian plate. Wallace sketched out a line that separated these regions, a line that is still called Wallace's Line. Wallace's Line marks the separation of placental mammals and marsupials as well as differences in birds and plants between these areas. It also separates West Malesia from East Malesia. The islands of West Malesia all lie upon the Sunda Shelf, a shallow continental shelf that was exposed during glacial maximums, so plants and animal could move freely among the islands. The islands on the Australian plate lie upon the Sahul Shelf. The two shelves are now only 15.5 mi (25 km) apart. Between them is a deep oceanic trench that forms a formidable barrier to dispersal.

begin in the eastern Pacific, affect this region. During an El Niño, the monsoon is weakened and pushed equatorward, creating a period of prolonged drought. Detrimental to the rainforest, it is devastating to the poor countries of the region because it causes widespread crop failure and food shortages. Some researchers argue that El Niños initiate the mass flowering events common in the Asian-Pacific rainforest.

Tropical cyclones (typhoons, hurricanes) are other important influences on the rainforest. The areas most affected lie between 10–20° N and S latitude and include Bangladesh, the Philippines, and much of Melanesia into Australia to the south. The cyclical incursions of cyclones into forested areas have left stands of disturbed forest where fast-growing, sun-loving pioneers tend to dominate. Other catastrophic events affecting these rainforests are earthquakes, landslides, volcanic activity, and tsunamis. Along with widespread destructive effects on the forest, these events have had disastrous consequences for the human populations of the region.

Soils

Soils of the Asian-Pacific rainforest are similar to those of other rainforests, but the percentage of area covered by each soil type differs significantly. While the deeply weathered and infertile oxisols made up a large percentage of Neotropical and African soils, in the Asian-Pacific region they account for only 3 percent. Oxisols occur in patches on the islands of Borneo, Sumatra, Java, Sulawesi, and the Philippines, and on the mainland in Thailand and Malaysia.

Ultisols are the most abundant soils in the Asian-Pacific rainforest. They occupy a significant portion of the forests in Malaysia, Sumatra, Borneo, Sulawesi, and the eastern Philippines. Like oxisols, ultisols are deeply weathered soils with low fertility. Ultisols are susceptible to erosion and movement due to the clay layer at depth.

Inceptisols and entisols cover equal amounts of land in the region. Three types of inceptisols are significant: aquepts (gleysols), andepts (andosols), and tropepts (cambisols). Although once forested, aquepts in tropical Asia are now devoted to rice production. They have high soil fertility and support large human populations. Andepts are of volcanic origin and exceptionally fertile. They are important soils in the Philippines, Indonesia, and New Guinea. Tropepts are well-drained nonvolcanic soils that occupy large areas of the Asian-Pacific region. They are similar to oxisols and ultisols in that they have low fertility and are reddish in color. Tropepts make up over half of the inceptisols found in the region.

The entisols are also divided into three groups: the fluvents (fluvisols), lithic group (lithosols), and the psamments (arenosols and regosols). A little over one-third of the entisols found in the Asian-Pacific rainforest are fluvents. These are well-drained, young, alluvial soils found along river valleys and subject to flooding. They are among the richest agricultural soils, and in Asia, they have been extensively converted for rice production. Another 40 percent of the entisols in the Asian-Pacific region are in the lithic group (lithosols). These are shallow soils on steep slopes, near rock outcrops. Psamments are deep, sandy soils that are acidic and low in fertility. Tropical heath forests, or kerangas, are found on psamments. In the Asian Pacific, these soils occur on Borneo and Sumatra.

The remaining soil types are histosols and alfisols, with a fraction of spodosols, aridisols, vertisols, and mollisols. Thirty-three percent of Asian soils are relatively fertile in contrast to the dominance of infertile soils in the Neotropical and African rainforests.

Vegetation

Floristically this region is called Malesia and is separated into two more or less distinctive subregions. West Malesia includes India, Southeast Asia, the Philippines, the Malayan Archipelago, Brunei, and Indonesia including Borneo. East Malesian includes the area east of Wallace's Line: Sulawesi, Lombok, New Guinea and nearby tropical islands, and northeastern Australia. (The term Malesia should not be confused with the country Malaysia.)

Forest Structure
The flora of the Asian-Pacific rainforest is rich and diverse; over half of the world's flowering plants families are represented. About 2,400 genera and 25,000–30,000 species are known, many of which are endemic, although the degree of endemism varies among the islands. Trees are mainly evergreen, with some semi-evergreen trees occurring on the fringes of the rainforest where it blends into the monsoonal

forests. Tree diversity is equal to or higher than the Neotropical and African rain-forests. One study in Borneo identified 300 different trees on a 2.4 ac (1 ha) plot.

The forest structure contains several canopy layers with intertwined vines and epiphytes (see Figure 3.16). Along with the shrub and surface layers, they create multiple layers of habitat for the creatures of the forest. The structure of the Asian-Pacific rainforest is slightly different from that of its counterparts. Although several distinctive canopy levels are apparent, the emergent layer is composed of clumped trees typically of one family and often one species. Many of the trees that make up the canopy layers are diptocarps (family Dipterocarpaceae). From a bird's-eye view, a distinctive pattern of groups of emergent trees rising above the canopy is quite evident, particularly in the West Malesian forests (see Figure 3.17). Emergent trees are very tall, typically 195–230 ft (60–70 m). Like emergent trees elsewhere, many have buttressed trunks.

The trees below the emergents, also dominated by dipterocarps in many areas, reach heights of 100–135 ft (30–41 m). Other trees present in the canopy are in the fig (Moraceae), laurel (Lauraceae), sapote (Sapotaceae), mahogany (Meliaceae), and legume (Fabaceae) families. The dipterocarp and leguminous trees prefer the sandier soils, while the other families grow on the less-fertile lateritic soils (oxisols and ultisols).

The plants in the understory of the Asian-Pacific rainforest include shrubs and saplings of canopy trees that are able to survive under limited light conditions. Palms are particularly abundant. Few plants grow on the forest floor, where herbs and ferns are present along with seedlings of the canopy trees. Some of the

Figure 3.16 Forest structure in an Asian rainforest includes tall dipterocarp species that tower above the canopy. *(Illustration by Jeff Dixon. Adapted from Richards 1996.)*

Figure 3.17 Rainforests of Sumatra stretch over the mountains but are rapidly being destroyed. Taken in Gunung Leuser National Park, Aceh Province, Sumatra, Indonesia. *(Photo courtesy of Gareth Bennett.)*

flowering plant families present are the gingers, sedges, aroids, African violets, and orchids. Fern-like spiked mosses occur in the understory and on the forest floor.

Dipterocarpaceae

The rainforests of West Malesia are like no others in the world because of the dominance of a single family, Dipterocarpaceae. The family name means "two-winged fruits," which describes the appearance of their fruits that come in many different sizes and shapes. Tens of genera and hundreds of species are found almost exclusively within this region. Dipterocarps have been present for a very long time. Dipterocarp pollen found in this area has been dated to 30 mya, a time coinciding with the collision of the Indian plate with Asia. Dipterocarps are thought to have originated in Africa and reached Southeast Asia via the migrating Indian plate. They then underwent a massive evolutionary radiation in Southeast Asia. The later collision of the Australasia plate with Asia may have allowed the dipterocarps to disperse into Sulawesi and the Pacific region. Dipterocarps dominate the forests of Borneo, Java, Sumatra, and the Malay Peninsula and the wetter forests in the Philippines.

Dipterocarps can be immense trees, exceeding 200 ft (62 m) in height; they protrude out of the canopy. They have smooth straight trunks without side branches until they reach the canopy. The base is often buttressed. In the canopy, these trees produce a crown shaped somewhat like a cauliflower. They rarely fall like the emergent trees of the Neotropics. Instead, they die standing in place. Dipterocarp trees produce an oily resin useful in the defense of bacteria, fungi, and animals. This resin accumulates wherever the bark is bruised. When hardened, it is called "dammar" and used by local people as a varnish or a boat caulking agent. The kapur or camphor tree, a dipterocarp common in Borneo, produces camphor, an essential oil used in medicines and as a preservative.

Flowers of the dipterocarps vary in size from small to large; they have five petals and many stamens and are scented to attract the thrips, beetles, moths, and bees that pollinate them. The fruit produced is a single-seeded nut with a wing-like covering that allows the seed to spin to the ground (see Figure 3.18). These winged seeds were probably helpful in the dispersal of this plant family throughout Southeast Asia and into New Guinea. Many dipterocarps share a regeneration strategy that involves gregarious flowering and consequent mass fruiting. This often occurs in two- to seven-year intervals. Flowering occurs at the same time in different species among multiple layers of the canopy. A succession of flowers occurs over about one month's time. The flowering is thought to be cued by brief episodes of cooler night temperatures or by drought (such as that during an El Niño year), although more research is needed on this subject. After flowering, an abundance of fruit appears, providing food for many resident forest animals as well as those that migrate to the area to feed. The seeds germinate quickly, forming a bed of seedlings on the forest floor. These seedlings start out tolerant of shade and may be able to survive several years under the dense canopy. Long-term seedling survival in the

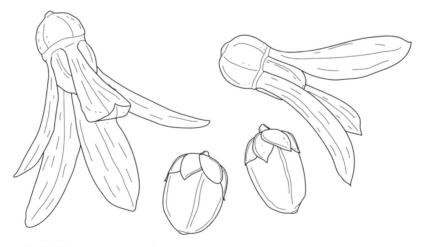

Figure 3.18 Dipterocarp seeds have wing-like appendages that allow them to disperse away from the parent tree. *(Illustration by Jeff Dixon.)*

shade may be made possible by the presence of a mutualistic relationship between the roots and mycorrhizal fungi that helps the plants absorb nutrients. Young trees become light demanding as they reach the canopy. Tree falls and gaps stimulate rapid plant growth. Such gaps produce clumps of trees that grow into and above the canopy.

Dipterocarps produce an extremely valuable wood, which is in great demand and has led to the exploitation of these forests. Their growth habit of tall branchless trunks and the clumping of trees allow for large-scale clear-cutting practices to extract them economically and efficiently.

The rainforests of the Asian Pacific region are relatively poor in climbing species apart from palms and dipterocarps. However, some climbers in the dogbane, milkweed, morning glory, legume, and squash families are present. Climbing palms are abundant and attain their greatest diversity in Southeast Asia. Spiny rattan palms are the most diverse group; as many as 30 species can coexist in the same area. These palms climb into the forest canopy with the help of hook-like spines on their young stems and the underside of leaves that allow them to grab onto the often-smooth bark of tropical trees. Rattan palms are in high demand for use in furniture and are intensively harvested. East Malesia is relatively poor in climbing plants other than palms and pandans.

Epiphytes are abundant throughout the Asian-Pacific rainforest except for Australia. Palms, ferns, orchids, and even dipterocarps are dominant. Orchids are found throughout Southeast Asian and New Guinea. Thirty-four percent of the world's orchids are found in tropical Asia and New Guinea, but only 3 percent occur in Australia. Species of the genus *Bulbophyllum* reach their peak in these rainforests. The flowers of these orchids are often tiny and purple in color and have a foul smelling odor to attract their main pollinator, flies. More than 1,000 kinds of large, showy dendrobium orchids are also found in this region. They seem to come in every possible color and shape. Bird's nest ferns are abundant and fill a niche similar to the bromeliads in the Neotropics. They trap dead leaves and organic matter, but they do not form water-filled reservoirs.

In East Malesia, dipterocarps are present but play a less-dominant role than in West Malesia. Trees in the laurel, melastome, nutmeg, and mulberry families are also found here. Many endemic species, including primitive conifers from the araucaria (Araucariacea) and podocarp (Podocarpaceae) families, relicts of ancient gymnosperm forests, are found in the rainforests of New Guinea and Australia. These conifer families show a disjunct Gondwanan distribution, meaning they occur in widely separated regions of South America, Australia, and New Guinea. They were probably more widespread prior to the breakup of Gondwana and the evolution of flowering plants.

Heath Forests. Heath forests are rainforests on sandy soils within Borneo. In Borneo, they are called "kerangas," a local term meaning "land where rice will not grow." The soils in heath forest are infertile. Kerangas have trees and shrubs with small leathery leaves. In forests with open canopies, the understory is dense. The trees are not buttressed and few climbing vines are present. The keranga tends to be less species rich than the lowland rainforest.

These heath forests are composed of dipterocarps, but the species are different from those in the rainforest. Many small-leaved plants in the myrtle family are present, as are casuarinas and conifers. Many of these forests tend to be dominated by a single or a few species. Kerangas form closed forests with the highest trees about 98–115 ft (30–35 m) tall. Some kerangas are pole forests, with densely packed straight thin trees 16–39 ft (5–12 m) tall; others are open woodlands.

Wet Forests. Mangrove forests are found throughout the Asian-Pacific rainforest along the coasts. They have a simpler structure, with a single canopy layer with individual species forming pure stands. Different species are distributed in different zones depending on water level, salinity, and tidal regime. They typically have no epiphytes or ferns.

Peat swamp forests are another forest type found in Southeast Asia. Peat forests grow on layers of slowly decomposing vegetation (peat) 3–65 ft (1–20 m) deep. Water enters the swamp through rainfall, but drainage is poor. Peat forests develop in concentric circles from the center (the wettest part) outward. In the center, the trees are quite small, away from the center trees get taller and eventually become lowland rainforest containing several dipterocarps wherever drainage is sufficient.

Freshwater swamps occur along rivers and can be temporarily or permanently covered with water. These swamps contain a mixed forest with several canopy layers. The tallest trees are less than 160 ft (50 m). Here epiphytes and ferns are more common.

Wet forests occupy 10–15 percent of the area of most Southeast Asian countries, but they are rapidly shrinking with drainage of the swamps and with development within mangrove forests.

Animals

The incredible diversity of flora within the Asian-Pacific rainforest provides abundant niches for a diversity of animals. Animals are able to exploit different canopy layers, with few moving between canopy and forest floor. Trees are home for most birds, as well as monkeys, squirrels, amphibians, reptiles, and myriad invertebrates in the Asian-Pacific rainforest. Many animals are able to glide through the forest. The forest floor houses larger herbivores and predators, along with smaller rodents, ground birds, reptiles, and amphibians. The separate designation of West and East Malesia is useful in discussing mammals. In the West Malesia subregion, placental mammals are dominant. Pangolins, monkeys, tapirs, elephants, rodents, otters, civets, leopards, and tigers among others are present. The area east of Wallace's Line is home to three species of monotremes and hundreds of marsupial species that occupy similar niches as placental mammals in the west. Bats are abundant in both areas.

Mammals

The different mammal families in the West Malesia subregion of Asian-Pacific rainforest are listed in Table 3.4. The Asian rainforest, like the African forest, has pangolins, but not the arboreal forms found in Africa. All Asian pangolins are terrestrial. They rest in burrows during the day and search the forest floor at night for ants and termites. The two species found in the Asian-Pacific rainforest are the Malayan pangolin and Indian pangolin.

Tree shrews are neither true shrews nor arboreal. They have been put in their own taxonomic order, Scandentia. Tree shrews probably evolved from insectivores, although some argue they descended from primitive primates. Resembling ground squirrels, they have bushy tails and narrow snouts. Tree shrews have the highest brain-to-body-mass ratio of any animal, even higher than humans. They are commonly found on the forest floor, where by day they feed on fruit, arthropods, and seeds. A few species live in the tropical rainforests in West Malesia; others inhabit the tropical seasonal forests. They are absent from New Guinea and Australia.

The Asian-Pacific region ranks second to the Neotropics in terms of diversity of bats. Bats are the most numerous and speciose mammal of this entire rainforest region. Both Megachiroptera and Microchiroptera are represented by many species. The largest are called flying foxes (because of their fox-like heads) and are fruit eaters. Elongated fingers support their 4 ft (1.2 m) wingspan. Flying foxes inhabit forest and swamps. Most roost on the branches of emergent trees. At dusk, they fly into the forest canopy to feed on ripe fruit.

Other Old World fruit bats range in size from 0.5 oz–3 lbs (15–1,500 g). They are most active at night, but have been seen flying during the day. They often make long flights moving out of an area when fruit is no longer available. The larger fruit bats are gregarious and roost in large groups, while the smaller fruit bats are solitary. Many of them locate food by smell. Fruit bats play an important part in

Gliders of the Forest

On the island of Borneo, the tropical rainforest hosts a variety of animals that have a unique adaptation to life in the trees. Snakes, lizards, frogs, giant squirrels, and colugos all glide through the forest canopy trees. These forests hold more than 30 types of gliding animals, more than any other location on Earth. Most gliders are nocturnal and have cryptic coloring that lets them blend into the canopy. They have evolved different strategies for gliding success such as skin flaps, webbed feet, and flattened bodies.

Why so many gliders? The structure of rainforest in Borneo is different from other rainforests in that the many large emergent trees are clumped together. The upper canopy is discontinuous and has fewer climbing vines and fewer connections between the trees that serve as arboreal highways. Gliding is an efficient mode of travel in these circumstances. Furthermore, since most of the trees in the rainforest of Borneo are dipterocarps that fruit infrequently, food can be sparse. Animals must be able to move quickly and efficiently through the forest to find food. By gliding, these animals do not have to make the long trip up and down the tree trunks. They can remain in the canopy out of the reach of most predators. Borneo's isolation over time has allowed many animals to develop this unique adaptation. If one cannot fly, surely gliding is the next best thing.

Table 3.4 Asian-Pacific Mammals Found in the Rainforest

Order	Family	Common Names
Monotremes (Prototheria)		
Monotremata	Tachyglossidae	Echidna
	Orinthornynchidae	Platypus
Marsupials (Metatheria)		
Dasyuromorphia	Dasyuridae	Antechinus, quolls
	Potoroidae	Rat kangaroos
Diprorodontia	Phalangeridae	Brushtail possums, cuscuses
	Macropodididae	Wallabies, tree kangaroos
Peramelemorphia	Peramelidae	Bandicoots
Placental Mammals (Eutheria)		
Artiodactyla	Suidae	Pigs
	Cervidae	Muntjacs, sambar, axis deer
	Boviidae	Buffalo, banteng, gaur
Carnivores	Felidae	Leopards and golden cats
	Herpestidae	Mongoose
	Mustelidae	Otters and ratels
	Viveridae	Civets and genets
	Ursidae	Sloth bear, sun bear
Chiroptera	Craseonycteridae	Bumblebee bat and hog-nosed bats
	Emballonuridae	Sac-winged bats, sheath-tailed bats, and relatives
	Hipposideridae	Leaf-nosed bats
	Megadermatidae	False vampire bats
	Molossidae	Free-tailed bats
	Nycteridae	Slit-faced bats
	Pteropodidae	Fruit bats
	Rhinolophidae	Horseshoe bats
	Rhinopomatidae	Mouse-tailed bats
	Vespertilionidae	Insect-eating bats
Dermoptera	Cynocephalidae	Flying lemurs
Perissodactyla	Tapiridae	Tapirs
	Rhinocerotidae	Rhinoceros
Pholidota	Manidae	Pangolins
Primates	Cercopithecidae	Macaques, langurs, leaf monkeys, proboscis
	Galagidae	Galagos
	Hominidae	Orangutan
	Hylobatidae	Gibbons, lesser apes
	Lorisidae	Lorises
Proboscidea	Elephantidae	Elephants
Rodents	Sciuridae	Squirrels
	Muridae	Old World rats and mice
Scandentia	Tupaiidae	Tree shrews
Soricomorpha	Scoricidae	Shrews

pollination and seed dispersal for a variety of rainforest plants. Old World bats as a group are threatened due to the loss of tropical rainforests. Others are killed because they are perceived as pests in orchards and plantations. Still others are hunted as food.

The majority of bats in the Asian-Pacific rainforests are small insectivores that use echolocation to find their prey. Many have distinctive noses and ears to aid in echolocation. Tomb bats, sheath-tailed bats, horseshoe bats, round leaf bats, wooly bats, bent-winged bats, bamboo bats, and lesser-yellow bats are found in these rainforests. Most are aerial insectivores, catching insects on the wing, but some are foliage gleaners. Others employ both strategies, taking insects from the air as well as off vegetation. Only a few fish-eating or otherwise carnivorous bats inhabit the area. From the endemic family Craseonycteridae comes the smallest known mammal in the world. With the common name of the bumblebee bat, this little bat is about the size of its namesake.

Australia's bats are similar to those of the rest of the region with the addition of the Australia false vampire bat or ghost bat, so-called because it has white or pale gray hair. They are the only carnivorous bats on the continent and eat large insects, reptiles, frogs, birds, small mammals, as well as other bats.

Rodents are represented by four families: squirrels, Old World porcupines, mice and rats, and bamboo rats. Asian squirrels and flying squirrels are members of the same family. Squirrels are particularly abundant in the West Malesia subregion of these rainforests. Most are arboreal, living in the upper and middle tree canopy and eating fruits, seeds, leaves, and insects. Squirrels are the dominant rodent of the Southeast Asian rainforest and occupy many niches taken by mice and rats in other regions. Squirrels vary in size from a few inches to rabbit size. Their fur differs in coloration from black, brown, or red, to white. Tree squirrels, ground squirrels, giant squirrels, flying squirrels, giant flying squirrels, and pygmy squirrels are abundant in the rainforest. Many medium-size tree squirrels and ground squirrels in the forest are insectivores. Most are active during the day. Giant squirrels are strictly arboreal, living high in the canopy. They have flattened tails to help them leap from tree to tree. Flying squirrels have large fur-covered flaps of skin between their limbs that allow them to glide from treetop to treetop. They cannot really fly, but are incredible gliders. Like other squirrels, they are also excellent climbers. Flying squirrels are nocturnal and have very large eyes so they can see at night. The pygmy flying squirrel is the smallest flying squirrel, less than 4 in (100 mm) in length. The red giant flying squirrel can grow to be more than 3 ft (0.9 m). It is the largest of the 14 species of flying squirrels within the canopy.

Old World porcupines are large, slow-moving animals that rely on their imposing quills rather than on speed or agility for defense. Some can weigh as much as 55 lbs (25 kg), while others weigh only a few pounds. Unlike New World porcupines, these animals are terrestrial and excellent diggers, living in burrows they construct. Their diets include many kinds of plant material, as well as carrion. Three genera of Old World porcupines are found in the rainforests of West

Malesia: the brush-tailed and crested porcupines are found throughout the region, while the long-tailed porcupines are endemic to Southeast Asia.

Rats and mice are less abundant. Fewer than 50 species of rats, mice, and bamboo rats occur in this region. Most are terrestrial, although many are excellent climbers. Mice tend to feed on fruits, seeds, and grasses, while rats also eat insects, mollusks, or crabs. The family of Old World mice and rats (Muridae) is the only one to have reached Australia and New Guinea. Forty-five species of rats and mice are found in New Guinea and five (three shared with New Guinea) are found in the Australian rainforest. Within these forests, marsupial counterparts to rats and mice occupy many of the available niches. Bamboo rats are a small family of rodents that live in the West Malesia subregion. They are fossorial but come to the surface to forage. Bamboo rats are most commonly found in bamboo forests and agricultural lands, eating roots and shoots.

Colugos are the single representative of the order Dermoptera. They are sometimes called "flying lemurs," because they can glide and look similar to lemurs even though they are not closely related. They are medium-size animals with a large, furry membrane that extends from the neck to tail. This membrane allows colugos to glide long distances with great maneuverability. Their feet have long, sharp claws for holding onto bark and branches. Colugos feed on fruit, young leaves, and flowers. They are nocturnal and sleep on tree branches or hollow trees during the day. They hang upside down while feeding and while traveling along branches. Colugos inhabit the West Malesian rainforests of Southeast Asia and the southern Philippines.

Primates are abundant in the West Malesian subregion of the Asian-Pacific rainforest but are absent in the East Malesian subregion. Lorises, tarsiers, gibbons, macaques, langurs, proboscis monkeys, leaf monkeys, and orangutans inhabit these rainforest. Lorises are slow-moving primitive primates from the rainforests of Asia. Slender lorises are found in the forests of India and Sri Lanka, and slow lorises are found in Southeastern Asia. Lorises are small, with thick fur, forward-facing eyes, and little to no tail. They are arboreal and grasp tree limbs with their hands and feet. They rest in tree hollows or on branches. Lorises are active at night and their diets include insects, shoots, young leaves, fruits, eggs, and small vertebrates. They consume all parts of their prey, including feathers, scales, and bones of the vertebrates, as well as the exoskeletons of insects.

Tarsiers are limited to Southeast Asia, eastern India, and some of the Philippine Islands. They are small primates with silky, buff, grayish-brown, or dark brown fur and round heads with large forward-directed eyes. Their muzzles are short, and they seem to have almost no neck. The hindlimbs are longer than their forelimbs, enabling them to make long leaps. They are nocturnal and strictly arboreal. During the day, they sleep in dense vegetation, usually on a vertical branch or in a hollow tree. They feed on insects and small vertebrates that they capture by leaping and quickly grabbing their prey with their hands. Tarsiers live in family groups or by themselves.

Plate I. Oxisols are red soils abundant in the tropical rainforest, Madidi National Park, Bolivia. *(Photo by author.)*

Plate II. Agricultural plantations are found where tropical rainforests once existed in Costa Rica. *(Photo by author.)*

Plate III. Tropical rainforest in Manuel Antonio National Park, Costa Rica. *(Photo by author.)*

Plate IV. Leaf insects have wings and legs that resemble leaves to avoid predation, such as these in Las Cruces, Costa Rica. *(Photo by author.)*

Plate V. Tropical flowers come in many shapes and sizes. They have evolved to attract specific pollinators. *(Photo by author.)*

Plate VI. Bromeliads are abundant in the trees of the Neotropical rainforests. *(Photo by author.)*

Plate VII. Three-toed sloths spend most of their time hanging upside down in tropical trees. *(Photo by author.)*

Plate VIII. The fer-de-lance is a deadly snake commonly found in the Neotropics. *(Photo by Snowleopard1. Courtesy of Shutterstock.)*

Plate IX. Poison dart frogs come in many bright colors which serve as a warning to would-be predators. *(Photo courtesy of David B. Smith, Ph.D.)*

Plate X. The okapi is a large herbivore of the African rainforest. *(Photo by Steffen Foerster Photography. Courtesy of Shutterstock.)*

Plate XI. The Great Blue Turaco is the largest member of the endemic turaco family in Africa. *(Photo by Steffen Foerster Photography. Courtesy of Shutterstock.)*

Plate XII. Macaques are common in the tropical rainforests of Asia. *(Photo by EML. Courtesy of Shutterstock.)*

Plate XIII. Spiny echidnas are egg-laying mammals that can be found in the tropical rainforests of New Guinea and Australia. *(Photo by Sandra Caldwell. Courtesy of Shutterstock.)*

Plate XIV. The great hornbill's crest is ivory and they are often hunted. *(Photo by Mike Blake. Courtesy of Shutterstock.)*

Plate XV. Variegated squirrels are common in the forest. This one is feeding on coconuts on the Guanacaste coast in Costa Rica. *(Photo by author.)*

Plate XVI. Bengal tiger populations are declining rapidly in the tropical seasonal forests of India. *(Photo by Craig Hansen. Courtesy of Shutterstock.)*

Primates in the Old World monkey and baboon family are found in abundance in the Asian-Pacific rainforest. The family is broken up into two groups, the cercopithecids that have cheek pouches and simple stomachs, and the colobines, which do not have cheek pouches and have more complex stomachs that give them the ability to survive on a diet of mainly leaves. Macaques are common cercopithecids. Most live in Southeast Asia and tropical India (see Plate XII). Macaques are primarily diurnal and can agilely move long distances through the trees, although they do spend time foraging on the forest floor. They tend to be vegetarians, feeding on fruits, berries, leaves, buds, seeds, flowers, and bark, although some will eat insects, eggs, and small vertebrates.

Langurs and leaf monkeys are diurnal, arboreal primates. They are small and slender with long tails and arms. Their fingers are strongly developed, but their thumbs are small. They have unique digestive systems that allow them to exist on a diet almost entirely of leaves. Langurs tend to be territorial and travel in groups of 5–15. They are often hunted for food, and several species are highly threatened or endangered.

Proboscis monkeys are famous for the males' long and pendulous noses. Proboscis monkeys are found on Borneo in the rainforest near water or in mangrove swamps. They are most active during the afternoon, eating mainly leaves, fruits, and flowers. Proboscis monkeys are the best swimmers among the primates. They are endangered due to forest destruction and the clearing of mangrove swamps.

Gibbons are common in West Malesian rainforests. Gibbons are similar to great apes in that they lack a tail and have similar dentition. However, gibbons are smaller and slender and have long arms and long canine teeth. They are considered "lesser apes." Gibbons swing from branches as they travel through the forest. They are upright walkers on tree limbs and on the ground. They typically walk with their arms held high for balance. Gibbons are mostly diurnal, are territorial, and travel in family groups. The crested, white-handed, dark-headed, silvery-capped, gray, white-brown, and Kloss's gibbons are small and extremely agile. They eat mostly fruit, but take leaves, eggs, and small vertebrates to supplement their diet.

The siamang is the largest gibbon. It lives in the higher elevations of the rainforests of Sumatra and Malaysia. Siamangs differ from other gibbons in size and appearance. They have black hair and are the most arboreal of all Asian-Pacific

Macaques

Long-tailed macaques are sometimes called crab-eating macaques, as they take crabs in the coastal forest. Local people consider this macaque terribly greedy, since it stuffs huge amounts of food into its cheek pouches. When no more food is available or no more can be ingested, it will punch the food into its gullet by pushing its cheeks with a clenched fist. Malaysians say these macaques are easy to catch. They will drill a small hole in a coconut and put a sweet banana inside. When the macaque discovers the banana, it reaches into the coconut to grab it. Refusing to let go of its prize, the monkey's hand is stuck inside of the coconut, making them easy to capture. Pig-tailed macaques are often caught and trained to pick coconuts. Once trained, the macaque will climb the tree and drop coconuts to its caretaker, who will collect them and bring them to market. A strong, mature macaque is able to harvest up to 700 coconuts in one day. Macaques are the favorite food of the Monkey Eagle in the Philippines.

primates. The first two digits of their feet are webbed—they are the only primate with webbed feet. Siamangs also have a goiter-like sac below their lower jaw, which can echo and amplify sounds. Their diet consists primarily of leaves and fruit, but they will also eat flowers, buds, and insects.

Only one great ape (Hominindae [Pongidae]), the orangutan, occurs in the Asian-Pacific rainforest, on the islands of Borneo and Sumatra. Each island has its own subspecies. Fossils and remains recovered in China, Vietnam, Laos, and Java suggest orangutans had a much wider distribution in the past. An adult male orangutan is about 4–5 ft (1.25–1.5 m) tall, and can weigh 65–100 lbs (30–50 kg). Females tend to be smaller. Their coat is rather shaggy, thin, and dark red or reddish-brown in color. Orangutans are primarily arboreal and active during the day. Like other great apes, they build a new nest each night in the trees. They rarely visit the forest floor. Their diet consists of mostly fruit, especially figs. They will also eat other vegetation, insects, eggs, and small vertebrates. Males tend to travel alone, while females travel and forage in small groups. The already limited range of the orangutan continues to shrink, largely because of hunting and forest destruction. In the past, many were taken for zoos worldwide. The species is highly endangered.

Elephants are the largest mammals in the rainforests of West Malesia. Asian elephants are smaller than African elephants with smaller ears and tusks. Unlike African elephants, females do not have tusks. The Asian elephant's trunk is smooth with only one terminal projection, compared with the African elephant whose trunk is telescopic with two terminal projections. In the rainforest, Asian elephants tend to be nocturnal and feed on bamboo, wild bananas, and other plants. They spend their days resting deep inside the forest.

The West Malesian rainforest has two families and three species of odd-toed ungulates—the Malaysian tapir and the Javan and Sumatran rhinoceros. The Malayan tapir is found in southern Myanmar, Thailand, the Malay Peninsula, and Sumatra. This distribution represents the remnant of a once-larger, perhaps prehistoric, worldwide distribution. It is easily distinguished from the Neotropical tapirs by its distinctive coat pattern: the front of the body and all four legs are black, and the back and rump are white. This pattern serves as camouflage and makes the defenseless tapir difficult to see in the shaded forest. Malayan tapirs can weight 600–1,000 lbs (272–453 kg). They tend to rest in the forest during the day, and come out at night to forage on grasses, leaves, shoots, and small branches near water. They are endangered due to deforestation and hunting.

Rhinoceroses were once widespread throughout Southeast Asia. Today, they are restricted to two large islands, each having its own species, the Sumatran and Javan rhino, respectively. The two-horned Sumatran rhinoceros probably still exists in the rainforest, but it is rarely seen. Sumatran rhinoceros tend to be solitary browsers, feeding on saplings, twigs, fruit and leaves at dawn and dusk. They have a prehensile lip used to grab and pull vegetation. Only 300 Sumatran rhinoceros are thought to remain in the highly fragmented rainforest. Many were killed and their horns collected illegally and sold to China, where the horns are ground up and used as an aphrodisiac.

The one-horned Javan rhino may already be extinct in the wild; the only known Javan rhinos are the 60 living in the Udjung Kulon Game Reserve in East Java and a few others in zoos. Javan rhinos are hairless with gray skin. Almost hunted to extinction in the 1930s, its current precipitous decline is due to loss of habitat and illegal poaching.

Four families of even-toed ungulates—the mouse deer, gaur, sambar and munt-jacs, and pigs—are also found in the West Malesian rainforests. Mouse deer or chevrotains are not really deer but are related to camels. They are very small, only 8–13 in (30–33 cm) at the shoulder. The mouse deer's face is similar to the South American agouti, and its legs are deer like. They do not have horns or antlers, but males have tusk-like upper canines. The mouse deer is native to the Malaysian Peninsula, Indonesia, and nearby islands. They are nocturnal ruminants and eat leaves, buds, shrubs, and fruit. Mouse deer are hunted for their skin, which is made into handbags and coats. They are also taken for their meat or are captured for pets. Mouse deer figure prominently in many Malaysian folktales.

The gaur is a wild ox from the rainforests of West Malesia. It is a large ox that stands 6 ft (1.8 m) at the shoulder. Gaurs have black bodies and whitish legs. Their massive curved horns have black tips. They move about in large herds within the forest and come to clearings at dawn and dusk to feed.

Two kinds of deer live in West Malesia: sambars and the barking deer or munt-jac. Sambars are large deer found in dense forests and open scrub. Sambars are nocturnal, solitary browsers. They are hunted for food and trade. Barking deer or muntjacs are smaller deer with small antlers. They are known as barking deer due to the sharp loud bark-like call they produce when alarmed.

Both the common wild pig and the bearded pig are found in the rainforest. The wild pig travels in large groups of 50 or more and is preyed upon by many forest carnivores. The bearded pig is larger and lighter in color. It is named for the bushy tufts of hair that grown on its snout. Bearded pigs can weigh several hundred pounds. They have dangerous long, curved tusks.

An abundance of prey supports many carnivores in the West Malesian subregion of the Asian-Pacific rainforest. Many civets, linsangs, mongooses, and a few species of otters, bears, and small and large cats are found in the forest. Civets and linsangs are medium-size cat-like carnivores with long bodies and short legs. Most species have small heads with long pointed muzzles and retractable claws. Many civets and linsangs are nocturnal hunters, hunting small vertebrates, insects, worms, crustaceans, and mollusks. Some are strictly carnivorous and others include fruit and roots in their diets. Many are strongly arboreal. Linsangs are solitary and nocturnal, spending most of their time in the trees.

The binturong is the largest arboreal civet. It has long, coarse, black hair and a long prehensile tail. The binturong lives in the trees of dense forest and is rarely seen. It can dive, swim, and catch fish or hunt for birds. It will also eat carrion, fruit, leaves, and shoots.

Mongooses are common predators in the West Malesian subregion. They resemble civets in appearance and distribution, but they tend to be smaller with more uniformly colored fur. Mongooses have long bodies, short limbs, and five toes on each foot. Their claws are not retractable. Most mongooses are diurnal and form large social groups. They are known for their ability to kill and eat snakes, including cobras.

The sun bear and sloth bear are the only members of the bear family found in the rainforests of West Malesia. The sun bear lives in Southeast Asia, Myanmar, Malaysia, Sumatra, Borneo, and the Assam region of India. It spends most of its time in trees eating lizards, birds, fruit, ants, termites, and honey. The sloth bear lives in southern India and Sri Lanka and feeds largely on termites and bees. Bear populations are in steep decline due to loss of habitat and hunting. Bear gallbladders are thought to have great medicinal value and are used throughout Asia. Bears are also hunted for their skins and meat.

Small, medium, and large cats are found in the West Malesian subregion. The leopard cat is the size of a domestic cat and has brown fur with lines of black spots on its back. The Asian golden cat is about twice the size of a leopard cat with a small head and long legs. Although mostly terrestrial, it climbs well and hunts day or night in search of birds and small ungulates. The Asian golden cat population is threatened throughout its range due to deforestation. The leopard cat is more adaptable to deforestation, but it is heavily hunted for the fur trade. The fishing cat is another small feline of India and Southeast Asia.

The clouded leopard lives in the forests from Southeast Asia to Java. It is the only species within its genus. The coat of the clouded leopard is yellow with dark markings that form circles and ovals (clouds). The forehead and tail are spotted. The clouded leopard is largely arboreal and hunts in trees or pounces on prey on the ground from the trees above. Prey includes birds, monkeys, pigs, goats, and deer. The clouded leopard is endangered due to loss of forested land. These cats have been excessively hunted for their valuable pelts, which are traded on the black market.

The largest of the Asian cats are leopards and tigers. The spotted leopard is a large cat with a very long tail. The spotted leopard's fur can be straw-colored and spotted or completely black (these are often called black panthers). Leopards are nocturnal, resting in trees during the day. Their diet varies from monkeys and ungulates to rodents, rabbits, and birds. Several subspecies of spotted leopard are endangered. Forest destruction, loss of prey, as well as hunting for trophy, or its beautiful fur, are responsible for the small number of spotted leopards alive today.

Indonesian and Sumatran tigers are still present in the West Malesian subregion of the Asian-Pacific rainforest, although numbers are dwindling. These tigers are smaller than their Siberian cousins, but they are the largest carnivore in the rainforest. Tigers are nocturnal, solitary hunters that are extremely powerful. They roam and defend large territories while searching for food. Tiger populations are severely affected by deforestation and loss of habitat, as well as illegal poaching and trade. Many are killed for their fur as well as other body parts, which in some Asian cultures are used in medicinal preparations. Tiger bones are thought to give strength, as well as relieve pain and arthritis. Other tiger parts are sold as aphrodisiacs. The Sumatran tiger is critically endangered, with less than 400 remaining. Tigers are no longer found on Bali and Java. Tigers are present in India and on the Asian mainland, but these populations are in steep decline.

With the exception of rodents and bats, the mammals of East Malesia are very different from those in the west. This subregion is home to two additional types of mammals: monotremes and marsupials. Monotremes are found nowhere else in the world. Unlike other mammals, monotremes lay eggs instead of giving birth to live young. Platypus and echidnas are the only monotremes in the world. The platypus is found only in Australia, mainly along its eastern coast including the rainforests of Queensland. The platypus spends most of its time in water, hunting for larvae and small invertebrates. They live in small burrows along the water's edge.

Echidnas are present in Australia and on the island of New Guinea. The short-beaked echidna is found in Australia, and the long-beaked echidna inhabits the forests of New Guinea (see Plate XIII). Echidnas are covered with long spines, which they use in defense. When threatened, the echidna can coil into a ball of spines. In soft soil, it will quickly bury itself. Echidnas are specialized feeders that eat only ants and termites, similar to pangolins and armadillos.

Marsupials dominate the mammalian fauna of the East Malesian subregion east of Wallace's Line and fill the same niches as placental mammals in West Malesia. Marsupial families include herbivores as well as insectivores and small carnivores. The long-tailed pygmy possum is one of the smallest of all possums. They have long prehensile tails and spend their life in the forest canopy, where they eat insects and collect nectar from flowers. Feather-tailed gliders can transverse as far as 65 ft (20 m) on one glide. Both of these small possums are nocturnal.

Ringtail possums and gliding possums are medium size and mostly arboreal. Ringtail possums have prehensile tails that they use as a fifth limb to balance on tree branches. The common ringtail, green ringtail, and lemuroid ringtail are some of those found in the rainforests. Like the placental gliders, marsupial gliders have a membrane between their limbs and body that allow them to glide through the forest. The sugar glider is the smallest glider, weighing only 4–6 oz (115–160 g). The greater glider is the largest, twice the length, and 10 times the weight of the sugar glider. Both ringtail possums and gliders eat leaves, flowers, and fruit. They will also take nectar, pollen, sap, and occasionally insects.

Brushtail possums and cuscuses make up another group that occurs in the canopy of the East Malesian rainforest. Although they are in the same family, they are quite different. Brushtail possums are more common in Australia and occur in many different forest types. Brushtails have a pointed snout, long ears, and hairy tail. They move rapidly through the forest. Cuscuses are more common in New Guinea and restricted to the rainforest. Cuscuses have a flat face, short ears, and a mostly naked, scaly, prehensile tail. They move slowly through the forest. Both are arboreal and great climbers. Both eat fruit and leaves. Cuscuses will also eat insects, eggs, and nestling birds. In New Guinea, cuscuses are hunted for their meat and soft dense fur.

Kangaroos and their relatives that live in the rainforest include pademelons, wallabies, and tree kangaroos. Pademelons are small kangaroo-like marsupials that prefer solitary nocturnal life in the rainforest. They eat fallen leaves and fruits. Forest wallabies are larger than pademelons and look like small kangaroos. They have small ears and long tails used for stability. Their hindquarters are more muscular and their hind legs are longer than their forelimbs. Dark in color they blend in well in the dark forest understory. Like pademelons, forest wallabies are mainly nocturnal, solitary browsers.

Tree kangaroos are the largest arboreal herbivores of the East Malesian rainforest (see Figure 3.19). They stand about 1.5–2 ft (450–650 mm) tall and have long, thick tails measuring 2–3 ft (600–900 mm) and weigh 8–29 lbs (3.7–13 kg) depending on the species. Tree kangaroos' legs are stouter and sturdier than those of their terrestrial cousins. They are agile climbers and can leap from branch to branch or to the ground. They are nocturnal and solitary. They browse on leaves and eat fruit and flowers. Because of their large size and arboreal lifestyle, the tree kangaroo has few predators other than the New Guinea Harpy Eagle and humans. They are hunted in New Guinea for their thick fur and meat. Clear-cutting also threatens their survival.

Rainforest bandicoots are rat-like marsupials with long pointed noses and large ears. This family only occurs on New Guinea and its nearby islands. Its four genera contain the spiny bandicoots, Seran Island bandicoot, New Guinea mouse bandicoot, and the New Guinea bandicoot. Rainforest bandicoots are terrestrial, nocturnal omnivores. They are an important protein source for the indigenous people of the rainforest.

Quolls and antechinus are the carnivorous marsupials of the rainforest. Quolls tend to be aggressive, nocturnal animals. They feed on birds, young rats, gliding possums, and other small arboreal and terrestrial mammals, reptiles, and invertebrates. The spotted-tail quoll is the largest marsupial carnivore in the rainforests of this subregion. It is 1–2.5 ft (380–760 mm) in length and weighs up to 15 lbs (7 kg). Antechinus are mouse-size carnivores that eat mostly invertebrates such as beetles, spiders, and cockroaches. They are often called marsupial mice. Brown antechinus males have a peculiar tendency to die near the end of or just after the mating season, when they are less than a year old. The stress of frantically looking for mates, mating for six hours at a time, and aggressively fighting other males, probably leaves them susceptible to parasites and infections that quickly kill them.

Figure 3.19 Tree kangaroos are the largest herbivores in the rainforests of Papua New Guinea. *(Photo by Susan Flashman. Courtesy of Shutterstock.)*

Birds

The distinction of West Malesia and East Malesia is useful in discussing birds, as the birds in each subregion are quite different. The rainforest birds of Asia and Africa are similar to each other at the family level. This is most likely due to dispersal during intermittent forested connections. Some of the important rainforest families shared between the two regions include the hornbills, bulbuls, and sunbirds.

Hornbills are large birds with black, white, and yellow plumage and huge bills. Many have developed a casque, an outgrowth that adorns the top mandible of the bill. The casque is hollow and composed of keratin. The casque provides an indication of the age, sex, and status of an individual bird. It is a prominent feature in many species and poorly developed in others (see Plate XIV). The rhinoceros hornbill is named for its red casque that resembles the horn of a rhinoceros. Hornbills have broad wings and loud calls that can resemble the calls of monkeys. Hornbills eat almost anything, though some are strictly carnivorous and others frugivores. Hornbills have a unique nesting behavior. When the female is about to lay, she finds a crevice or hole in a tree and seals herself inside. Often the male will assist her. The hole is completely sealed except for a small slit, which is used by the male to pass food into his mate and their chicks. While in the nest, the female molts completely. She remains sequestered through incubation and the initial rearing of the young. The male, often helped by last year's brood, brings fruits, berries, and insects throughout the entire nesting cycle. Because hornbills are large birds, they require large expanses of forest. Along with most of West Malesia, hornbills are found in New Guinea but not Australia. As the forest declines, the populations become threatened.

Parrots are poorly represented in West Malesia, but are plentiful in East Malesia. Only three small parrots (the blue-rumped, vernal-hanging, and the blue-crown hanging parrot) along with several parakeets are present in the rainforests of West Malesia. Many have a vibrant green plumage and brightly colored heads and breasts.

Pheasants, partridges, tragopans, firebacks, great and crested argus, and peacock pheasants are some of the many ground-dwelling birds that live in the forests of West Malesia. Doves and an astounding array of pigeons, including green pigeons, are abundant on the forest floor and the lower canopy levels. Pittas and broadbills are small, round ground dwellers with short tails and long legs that inhabit the undergrowth. Many pittas have blue, green, or red feathers. Broadbills are less colorful with brown or black feathers with yellow markings.

Myna birds are common in West Malesia. Mynas are dark brown, black, or gray birds with yellow bills and yellow feet. They all have a white patch on the underside of the wings. Mynas mimic the calls of other animals in the forest.

Sunbirds, flowerpeckers, spiderhunters, and honeyeaters are small, brightly colored birds. Like the sunbirds of Africa, Asian sunbirds are mainly nectar feeders with long curved bills. They are bright yellow, red, purple, or olive green. Several have a metallic sheen. Flowerpeckers are very small and live in the upper canopy. Spiderhunters are small, mostly yellow birds with large curved beaks used to pick spiders out of holes. Honeyeaters are the most abundant of the nectar-feeding birds in East Malesia.

Like Africa, little brown insect-eating birds are the most abundant birds in the Asian-Pacific rainforest. They tend to travel in mixed flocks composed of many of the same bird families found in Africa. Bulbuls, Old World Warblers, thrushes, flycatchers, shrikes, and a large variety of babblers may be members of a single flock. Babblers and Old World Warblers are well represented in Asia, with each group having more than 100 species. The babblers are the most diverse bird family in tropical Asia. They tend to be small and drab brown in color. Different species search for insects on leaves, twigs, or tree trunks in the lower canopy of the rainforest.

Other forest birds include starlings, robins, drongos, woodpeckers, piculets, and barbets. This region also has an extraordinary number of small and large cuckoos and bee eaters that feast on the abundant insects in the forest. A large number of small and large kingfishers live along the rivers, streams, and flooded forests in both the West and East Malesian subregions.

Leafbirds are a small group of closely related arboreal birds found throughout tropical Asia. Five are widespread in Southeast Asia, but a few are endemic to specific islands. Leafbirds come in various combinations of green, yellow, and sometimes a little blue.

The predatory birds of the forest include eagles, buzzards, hawks, eagles, kites, and falconets. All are excellent hunters with diversified diets of fish, reptiles, small rodents, and other small mammals. The Monkey Eagle of the Philippines is more than 3 ft (1 m) tall and lives high in the forest canopy. Its favorite food is macaques. Because of the loss of rainforest throughout the Philippines, about 200 remain. The New Guinea Harpy Eagle hunts tree kangaroos and other arboreal marsupials. Owls and nightjars are the avian predators of the night.

In New Guinea and Australia, birds have diverged greatly in appearance and behavior. The bowerbirds, honeyeaters, and birds of paradise represent a radiation of passerine birds on the once-connected islands. These areas also house a diversity of two widespread families, parrots and pigeons, as well as a few Old World families, like hornbills, sunbirds, white eyes, and starlings that are more recent arrivals.

Lyrebirds, bowerbirds, megapodes, birds of paradise, and cassowaries are endemic families of the East Malesian rainforest. Lyrebirds are ground-dwelling birds in the rainforests of Australia and New Guinea known for their incredible mimicking abilities. They can mimic birds, flocks of birds, and animals, as well as human and mechanical noises. Males have large fan-like tails used in mating displays along with their elaborate vocal performances. Bowerbirds are pigeon-size ground-dwelling birds that also have elaborate mating displays. Instead of a flashy display of tail feathers like the lyrebird, the male bowerbird builds a nest or bower, which he decorates with twigs, grass, colorful feathers, flowers, and berries. He sings in front of the bower hoping to attract mating females. Eighteen different species of bowerbirds live in Australia and New Guinea. Each has its own preferred decor, perhaps a mossy lawn or a display of flowers and pebbles.

The megapode family includes 22 species found exclusively in East Malesia. Megapodes are turkey-like in appearance with small heads and large feet. They are also called mound builders or incubator birds because of their unique nesting practice. Megapodes do not sit on their eggs; instead, they build mounds of forest litter under which they bury their eggs. These mounds provide heat from the decomposing vegetation to incubate their eggs. Other megapodes lay eggs in burrows, allowing geothermal activity or solar radiation to incubate their eggs.

Megapode chicks kick their way out of their eggs and emerge with a full set of feathers.

Birds of paradise are so named because their tail feathers resemble the flower of the plant with the same name. Forty-two different bird of paradise species inhabit the East Malesian rainforests. They vary in size (robin to crow size) and color. They are best known for the elaborate feather displays males perform during courting. Many have elongated feathers on wings, body, head, and tails that they shake and ruffle in front of females. Local people and collectors have hunted these birds for centuries, many to near extinction. Their beautiful feathers are used in tribal clothing and ceremonial headwear. Excessive hunting, as well as forest destruction, continues to decrease populations.

The flightless cassowary is the largest ground-dwelling rainforest bird in the world. (The ostrich of the African savanna is the only flightless bird or ratite living today that is larger.) The cassowary is native to the forests of New Guinea and Australia. They have long black feathers and a distinctive blue neck and head. Cassowaries have brightly colored (usually red) skin flaps or wattles under the neck that may serve to attract mates in the dark forest. They have a large keratinous casque on their head that they use to shovel and search for food on the ground. Like hornbills, the casque may indicate dominance and age. Cassowaries have powerful legs and feet that enable them to run at speeds up to 30 mph (48 kph). Their feet are equipped with sharp claws and the inner toe forms a long dagger able to rip through flesh. They are mainly frugivores, eating fallen fruit or fruit still hanging on branches, but they also consume small vertebrates, fungi, and insects. Cassowaries are important seed dispersers in the forest of New Guinea and Australia. Although populations of cassowaries are stable, they are vulnerable to forest clearing and hunting.

Other birds, such as fairy-wrens, honeyeaters, logrunners, Australian babblers, and warblers have species endemic to the East Malesia subregion of the Asian-Pacific rainforest.

Reptiles and Amphibians

Snakes, lizards, crocodiles, turtles, and a vast array of frogs, toads, and caecilians inhabit the rainforest. The warm, wet environment is quite suitable for their cold-blooded physiology.

More than a hundred different kinds of tropical snakes are found in the Asian-Pacific rainforest. Less than 10 percent are poisonous, and only a few of these are dangerous to humans. Poisonous snakes in this region belong to the Elapidae (cobras, kraits, and coral snake) and Viperidae (vipers, pit vipers, and adders) families. Elapids have short, hollow poison fangs in the front part of their upper jaw, while vipers have longer hollow fangs at the back of the upper jaw.

Two cobras, the king cobra and the Indian cobra, are found in the West Malesia subregion. Cobras expand their neck ribs to form a hood. In spitting cobras, such as the king cobra, the fangs face forward. The king cobra is the most deadly and the largest of the world's cobras, growing to lengths of about 13 ft (4 m). A few have been recorded close to 18 ft (5.5 m) long. Its diet consists of cold-blooded animals, mostly other snakes. The king cobra has an interesting brooding process. The female will lay eggs and keep vigil over them in the nest, while the male stands guard outside the nest. Both aggressively protect their eggs. The venom of the king cobra is an extremely potent neurotoxin that affects the nervous and respiratory systems.

The Indian cobra is a medium-size snake growing 6–7 ft (1.8–2.2 m) in length. The snake's coloring varies from black to dark brown to a cream, usually covered with a spectacled pattern. A wide black band on the underside of the neck and the hood markings of half-rings look like large eyes when viewed from behind and distinguish it from other cobras. The Indian cobra feeds on rodents, lizards, and frogs. Its venom damages the nervous system of the prey, paralyzing and often killing it. The Indian cobra is the snake often kept by the "snake charmers" of India. Although the snake appears to be dancing to the pipe music of the charmer, snakes cannot hear. It is actually provoked into a striking position and it sways as its gaze follows the charmer's hands and pipe. The result is the cobra's "dance." Several subspecies of the Indian cobra inhabit West Malesia; one spits venom into its victim's eyes and blinds it.

Kraits are another elapid. They are nocturnal and tend to be passive until provoked. Their venom is many times more potent than that of a cobra and quickly induces muscle paralysis. Chances of survival are only 50 percent even when anti-venom is administered. Kraits eat other snakes and small lizards. Several other venomous snakes can be found in the Australian rainforest, including two elapids, the red-bellied black snake, the highly venomous eastern brown snake, as well as the less-venomous brown tree snake.

Vipers, pit vipers, and adders are among the most dangerous snakes in the world. Many are common in the Asian-Pacific rainforest. Vipers' long fangs can be folded back in their mouth when not being used. Pit vipers usually give birth to live young, although a few species lay eggs. Most vipers have hemotoxic venom that affects the bloodstream, causing necrosis and eventual death if left untreated. Some of those in the rainforest include the Malayan pit viper, hundred pace viper, hump-nose and palm vipers, and the South Asian, bamboo, and temple pit vipers.

Nonpoisonous snakes are abundant in the Asian-Pacific rainforest, among them several pythons. The reticulated python is one of the largest snakes in the world. It is a constrictor whose main diet is birds and small mammals, such as monkeys. The largest recorded Malaysia reticulated python was 30 ft (9 m) in length and weighed 280 lbs (127 kg). Several other pythons include the Indian python and green tree python of New Guinea and Australia. Australia is also home to the carpet python and amethystine python. The amethystine python is a large snake, typically about 16 ft (5 m) in length. The record is 28 ft (8.5 m) long.

Tree snakes are smaller, often beautiful snakes that live in the trees eating birds, eggs, small arboreal mammals, and reptiles. They are fast and expert climbers.

Their coloration, similar to leaves and bark, provides good camouflage. The paradise tree snake of Borneo is a brilliant green color. It is often called a flying snake, as it is able to flatten its body and glide from tree to tree or across small rivers. Other large tree snakes are called cat snakes because they have vertically oval pupils, like those of a cat. A number of mildly venomous whip snakes are also found in this region.

Ground snakes include racers, rat snakes, keel backs, pipe snakes, and burrowing blind or worm snakes. Kukri snakes, reed snakes, little brown snakes, slug snakes, wolf snakes, and mock vipers are other common forest dwellers in the Asian-Pacific rainforest.

More than 148 species of lizards inhabit the rainforest, ranging in size from a few inches to 8 ft (2.4 m) giants. Lizards, geckos, and skinks are present in both subregions of the rainforest. Agama lizards are common and many different ones—including earless lizards; green-crested, forest-crested, and changeable lizards; and the Borneo anglehead agama and flying dragons—inhabit the rainforest. The flying dragon or Draco lizard is brightly colored and moves about the trees by gliding. It has skin flaps that fan out like an umbrella as it leaps from tree to tree. Several other gliding lizards—including the black-bearded, spotted, five-banded, Sulawesi, Blanford, and common gliding lizards—make the rainforest their home.

The largest lizards of the forest are the monitor lizards (the varanids). They are an ancient group found in Africa, central and southern Asia, Malaysia, the Indonesian islands, Papua New Guinea, and Australia. Monitors are strong, diurnal reptiles with long necks and tails. They range in length and weight from the short-tailed monitor at about 8 in (200 mm) and 0.7 oz (20 g) to the Komodo dragon at 10 ft (3 m) and 120 lb (54 kg). The Komodo dragon is the world's largest lizard. It is restricted to the islands of Komodo and Flores in Indonesia. Monitors have a varied diet, with small monitors taking fruit and mollusks, and the larger monitors attacking and killing large mammals such as deer. Large monitors also eat carrion.

Golden geckos, house geckos, Indo-Pacific geckos, tree geckos, flying geckos (Ptychozon), and Tokay geckos all inhabit this rainforest region. Tokay geckos, with a length of around 14 in (350 mm), are one of the largest geckos alive today. They live in trees and on cliffs in the tropical rainforests of northeast India and the Asian-Pacific region. The flying gecko is named for its ability to glide from tree to tree within the forest canopy. Flying geckos launch from branches while the air pressure flattens their body, limbs, and tail, extending the surface area as it glides.

Skinks are the most diverse group of lizards worldwide, and the greatest numbers of them occur in the Asian-Pacific region. They are slender and fast with pointed snouts and small limbs. Their movements resemble those of snakes. Skinks are carnivorous and eat invertebrates and small rodents. Tree skinks, sun skinks, brown skinks, slender skinks, and mangrove skinks among others call the Asian-Pacific rainforest home.

A number of tortoises and turtles live within the forest or along rivers. The Burmese brown tortoise and spiny hill tortoise are forest dwellers. The black pond tortoise and Malaysian box tortoise live in the swamps, ponds, and streams within the forest. The river tortoise spends most of its time in rivers. Like sea turtles, the river tortoise will come onto a river beach to lay its eggs. It will dig a hole, lay its eggs, and cover the hole as well as its tracks to avoid egg predation by mongooses that will cruise the beaches in search of the eggs. A few freshwater turtles, but no land turtles, live in the rainforests of New Guinea and Australia.

Toads, frogs, and caecilians are the amphibians of the Asian-Pacific rainforest. Many live in or near water, while others spend their lives in the trees or forest floor. True toads (Bufonids) are found all over the world. They have squat bodies, rough textured skin, and short legs. They tend to walk rather than hop. Some of the toads present in the Asian-Pacific rainforest are Sulawesi toads, Asiatic toads, forest toads, and four-ridge toads. True frogs (Ranids) are abundant throughout the region; they include crab-eating frogs, swamp, field and creek frogs, puddle frogs, cricket frogs, rock frogs, the Malaysia frog, and rhinoceros frog, among others. Litter frogs and horned frogs live on the forest floor among the decomposing leaves, and bullfrogs, chorus frogs, black-spotted, stick, and narrow-mouthed frogs (microhylids) emerge from burrows after it rains. Tree frogs are also abundant in the canopy and include Wallace's flying frog. Like the gliding snakes, lizards, squirrels, and lemurs, Wallace's flying frog glides through the rainforest. Their loose skin flaps and webbed fingers and toes provide them with the lift they need for gliding. These frogs can change their direction in midair.

Caecilians are legless amphibians that resemble worms or snakes. They spend most of the time in burrows in the tropical forests. They are mostly blind and have sensitive tentacle-like organs on their snouts that help them navigate and find prey. Caecilians eat worms and insects. Caecilians have not been well studied and much of their ecology and evolutionary history is still unknown.

Insects and Other Invertebrates

Like other tropical rainforests, the Asian-Pacific rainforest has a multitude of insects and other invertebrates that play important roles in the forest. Insects are the largest class of invertebrates in the rainforest. Butterflies, moths, ants, wasps, bees, termites, beetles, and stick and leaf insects are incredibly varied and have many unique adaptations to life in the forest.

Butterflies are abundant and colorful sights in the rainforest during certain times of the year. Many in this region are members of the five main butterfly families: (1) the birdwings and swallowtails, (2) milkweed butterflies, (3) gossamer-winged butterflies, (4) satyrs, and wood and tree nymphs, and (5) saturns and jungle glories. New butterflies are continuously discovered and identified. The birdwings and swallowtails are some of the largest and most spectacular butterflies of the region. They have long, pointed, bird-like forewings and tails on their hind wings. Birdwings tend to be large and brilliantly colored; their wingspans can reach up to 7 in (180 mm). Borneo alone has 11,000 species from this one family. The gossamer-winged butterflies are mostly small, brilliantly colored insects with long filamentous tails. They include the blues and hairstreak butterflies.

Moths are more numerous than butterflies. The largest moth, the Atlas moth has a wingspan of nearly 10 in (254 mm). Most moths are active at night; however, swallowtail moths are active during the day. Other moths include hawk moths, sphinx moths, and hornworms. Hawk moths are medium- to large-size moths. They have narrow wings and thin abdomens adapted for rapid flight. These moths are some of the fastest-flying insects, capable of flying at more than 30 mph (50 kph). They have wingspans of 1.5–6 in (35–150 mm). Hawk moths are noted for their flying ability, especially their ability to move rapidly from side to side while hovering.

Stick and leaf insects are abundant in the Asian-Pacific rainforest. Stick insects have long bodies and small limbs and closely resemble twigs. Most stick insects can fly, although their wings remain tightly folded over their abdomen and invisible when at rest. The largest stick insect measured, with a length of more than 12 in (300 mm), is from Borneo. Leaf insects are fatter and flatter than stick insects and look like leaves. Their bodies range from green to brown. Their legs are flattened, with leaf-like lobes that blend in with tree leaves. Stick and leaf insects are primarily nocturnal. They avoid detection during the day by remaining still. Although immobile and silent during the day, they are active at night as they feed on leaves.

Mantids also resemble leaves and sticks but for different reasons. While stick insects use camouflage to avoid predation, the mantids use camouflage to avoid detection by the prey they seek to capture. They lie in wait to ambush an unsuspecting insect. The orchid mantis's coloration and appearance blends in with the flower petals of a dendrobium orchid, where it lies in wait for any unsuspecting butterfly that happens to visit the flower. Other mantids look like decaying leaves. The dead-leaf mantis rests among the leaf litter on the forest floor until it spots its desired victim.

As in other rainforests, termites are abundant throughout the Asian Pacific and are the major decomposers of the rainforests. Damp wood termites feed mainly in fallen trees. This family of termites probably originated in this region. The majority of termites in the forest are in the higher termite family. Both soil-feeding and wood-feeding termites belong to this family. Most live in nests on the forest floor or underground, although some make nests in the trees. A few species are called processional termites and travel in ant-like columns carrying their food the way leaf-cutter ants do. Estimates of termite abundance exceed 1,000 individuals per square yard. In addition to playing a major role in the decomposition of fallen wood and recycling of nutrients, termites are a major source of food for pangolins, echidnas, shrews, and sun bears and sloth bears.

The Asian-Pacific rainforest has numerous ant species. In fact, on Mount Kinabalu on the island of Borneo, 640 different species of ants have been identified in just a few acres of forest. Red tree ants and fire ants are fierce defenders of the nest. They will quickly swarm an intruder delivering powerful bites. The giant forest ant is less harmful but quite large, about 1 in (25 mm) long. It nests in rotting trunks of fallen trees, and at night, travels high into the canopy to forage on honeydew, which accounts for about 90 percent of their diet.

Many bees, wasps, and hornets live in the forest, nesting high in the trees as well as low in the shrubs and bushes of the lower canopy. Social bees play a major role in the pollination of many dipterocarps in Southeast Asia. Larger carpenter bees also play a role in pollination. Bees in the Asian-Pacific rainforest must be able to adapt to long intervals between the massive flowering events of the dipterocarps. That may be why fewer bees are found there than in the Neotropics.

The Asian-Pacific rainforest is alive with the loud, sometimes screaming sounds of cicadas. Cicadas are sap-sucking insects. They are often difficult to see, but around dusk, their presence is certainly made known. The males call to attract females and the loud disharmonic chorus is hard to miss. Cicadas are the loudest insect in the rainforest.

Beetles are incredibly common and diverse within the rainforest. Some are specific to a given island, while others are found throughout the region. Jewel beetles are metallic green. Click beetles seem to throw themselves into the air to distract predators. Dung beetles are great decomposers of animal waste. Male rhinoceros beetles have elaborate horns used to fight other males and to acquire females. Two-horned and three-horned rhinoceros beetles occur in the Asian rainforest. Long-horn beetles have extremely long antennae. They lay their eggs in trees where the larvae bore a labyrinth of tunnels through the tree for several years until they emerge as adults. Weevils and fireflies are the abundant beetles in the forest.

Other Invertebrates. Spiders in the Asian-Pacific rainforests include classic web weavers, trap-door spiders, and those that sit and wait to ambush an unsuspecting victim. Orb web spiders build elaborate golden webs between trees or shrubs to catch large insects. To do so, the web must be very strong. In fact, the silk of the Nephila spider is the strongest there is, with a tensile strength twice that of steel. Very large tarantulas, such as the bird-eating spider that waits in a hole or crevice to ambush large insects, also find a home in the rainforest.

Scorpions are active hunters at night and spend their days under stones and bark or in rotting wood. They attack large insects. The Asian forest scorpion is among the world's largest, with a body and tail length of 6 in (152 mm). Whip scorpions are related to scorpions and spiders. They do not possess a stinger like true scorpions, but they produce an acidic mixture (mostly acetic acid) that they will discharge from their back end when disturbed or threatened by small rodents or other potential predators. Whip scorpions feed on worms, slugs, and other arthropods.

Centipedes and millipedes are both common forest creatures. Centipedes are nocturnal predators that feed on other invertebrates. They have powerful jaws that inflict a painful bite. They also use poison to subdue their prey. Millipedes are generally active during the day and feed on soft decomposing plant matter. Some millipedes can be quite long. One species in Borneo has been measured at 8 in (200 mm) long.

A rainforest would not be complete without leeches. They are prevalent in the lowland rainforest, where they hang from the leaves or twigs waiting to drop on an unsuspecting host. When it finds a host, a leech injects an anesthetic and an anticoagulant into the bite, so the victim does not feel the leech attach itself. The anticoagulant allows the blood to flow. The leech gorges itself and then drops off. The tiger leech does not have an anesthetic, so its bite hurts. Most leeches are up to about 2 in (60 mm) long, but the Kinabalu giant leech of Borneo can reach 12 in (300 mm) in length. Fortunately for most animals, including people, it feeds exclusively on the blood of large earthworms.

Human Impact on the Asian-Pacific Rainforest

The animals and plants of the Asian-Pacific rainforest have evolved together under the influence of the physical environment. Changes in rainfall or temperature can greatly affect their chances for survival. The destruction of the forest for forest products or conversion to agricultural land exacerbates the situation that tropical rainforests faced with global climate changes. With continuing growth of the human population forcing people to migrate into the rainforest and practices such as unsustainable forestry and illegal poaching of trees and animals, the Asian-Pacific rainforest is in jeopardy.

Asian-Pacific rainforests are the main source of tropical hardwoods for the global market. Many areas, once cleared of their trees, are converted to agricultural land for food and plantation crops such as oil palm and rubber.

Rainforest destruction has accelerated in the last few decades, mostly due to unsustainable timber extraction and agricultural conversion. In many areas, the primary forest has been completely lost. The rainforests of Vietnam are down to 20 percent of their original acreage. In Indonesia, government-sponsored migration into less-populated forested areas has led to the destruction of rainforest for the creation of towns and agricultural lands. Agricultural attempts on infertile forest soils often have led to crop failure and abandonment of the area.

Explosive population growth and the need for land as well as revenue have led to the massive exportation of timber. Unsustainable and illegal logging throughout Malaysia, Indonesia, Sri Lanka, and Myanmar has contributed to rainforest loss. Civil unrest is an additional factor in forest destruction in some of these areas.

Although Borneo has some intact rainforests, population pressure and timber interests have recently encroached upon the last large remnants of primary forest, leaving their future existence questionable. Intense pressure to clear forests for its valuable timber as well as to plant oil palm plantations pose serious threats to the remaining forests.

....

Palm Oil and the Loss of Rainforests

The oil palms (*Elaeis*) of West Africa have been imported worldwide to produce palm oil for grow-ing markets. Palm oil is used as a biofuel, ingredient in food products, cosmetic base, and engine lubricant. Oil palm productivity is high, yielding more per acre than any other tropical crop and provides strong economic growth in Malaysia and Indonesia. Unfortunately, oil palm production involves clearing large expanses of rainforest and the heavy use of pesticides and herbicides. Once the forest is cleared, oil palms are planted and can produce for about 25 years. Then the plants become too tall to collect the fruit (from which the oil is derived) economically. After that time, the land is deserted and turns to scrub vegetation. Clearing the forest for oil palm production leads to the extinctions of many species. Indonesia has developed plans to create oil palm planta-tions in and around national parks and reserves, fragmenting the only large remnants of undis-turbed rainforest in the region.

Economic development created by forest clearing for timber and the cultivation of oil palms is not sustainable in most areas. Demand for palm oil by China largely fuels this development. Indo-nesia and Malaysia continue to expand their economies and have found palm oil production to be highly profitable; unfortunately, however, the loss of rainforest and biodiversity is extremely high.

Proposals from industrial countries to pay for rainforest conservation in exchange for carbon credits to offset their carbon emissions may provide limited relief. Demand for sustainable and re-sponsible production of oil palm from importing nations may help to limit some deforestation, but many have shown reluctance in requiring responsible production.

....

The rainforests of New Guinea are rapidly being exploited. The western, Indo-nesian side of New Guinea (Papua and Iran Jaya) is experiencing rapid destruction as the Indonesian government continues to settle people on the island. On the east-ern side, in Papua New Guinea, agricultural conversions and timber production once occurred on a small portion of the area, but these conversions have been expanding rapidly since the year 2000. Increased population, timber, and mining explorations continue to be threats. The rainforests of Australia remain largely intact and protected, although they are extremely vulnerable to introductions of invasive species. Many of the rainforests of the Asian Pacific are vulnerable to introductions of invasive species. Humans have brought many species (intention-ally and unintentionally) into areas where these populations can explode due to the lack of natural predators. Islands are particularly vulnerable when these introduc-tions occur and have wiped out entire populations of endemic species.

Protected areas and reserves are found throughout the region. Many are there to protect individual species such as the Asian elephant, rhino, orangutan, and Komodo dragon, while others protect entire forests. In late 2007, Indonesia, with financial assistance from Australia, set aside 173,000 ac (70,000 ha) of peat forest in Borneo. The burning of peat forests is a large contributor to greenhouse gas emissions. By preserving this area, Indonesia is conserving biodiversity as well as limiting greenhouse gas emissions in this part of the world.

The Asian-Pacific rainforest contributes greatly to global biodiversity. Continued efforts by the countries of this region, working with local and international environmental organizations, can save much of the remaining rainforest. Like the African region, sustainable development, sustainable forestry, and ecotourism will help support the people and conserve the tropical rainforests of the Asian Pacific now and in the future.

Each of the regional expressions of the Tropical Rainforest Biome shows similarities in forest form and function. Some plant families like the legume or fig families are found throughout the biome, while others, such as the dipterocarps are restricted to one region. This is true of vertebrates and invertebrates as well. Similarities as well as distinct differences are evident. All three regions suffer from extensive deforestation and forest fragmentation due to human encroachment and exploitation. Each region's unique assemblage of biodiversity remains in jeopardy with the continued destruction of the forests. Local, governmental, and international groups are working diligently to protect some of the remaining forests.

Further Readings

Books

Aiken, S. R., and Colin H. Leigh. 1992. *Vanishing Rain Forests: The Ecological Transition in Malaysia*. Oxford: Clarendon Press.

Garbutt, N. 2006. *Wild Borneo: The Wildlife, Scenery of Sabah, Sarawak, Brunei and Kalimantan*. Cambridge, MA: MIT Press.

Gentry, A. H., ed. 1990. *Four Tropical Rainforests*. New Haven, CT: Yale University Press.

Kircher, John. 1997. *A Neotropial Companion*. Princeton, NJ: Princeton University Press.

Martin, C. 1991. *The Rainforests of West Africa: Ecology, Threats and Conservation*. Basel: Birkhäuser Verlag.

Primack, R., and Thomas E. Lovejoy, eds. 1995. *Ecology, Conservation and Management of Southeast Asian Rainforests*. New Haven, CT: Yale University Press.

Shuttleworth, C. 1981. *Malaysia's Green and Timeless World*. Singapore: Heinemann Educational Books.

Yamada, I. 1997. *Tropical Rain Forests of Southeast Asia: A Forest Ecologist's View*. Honolulu: University of Hawaii Press.

Internet Source

Butler, Rhett A. 2007. Tropical Rainforests. Mongabay.com. http://rainforests.mongabay.com.

Appendix

Common and Scientific Names of Species in the Tropical Rainforest Biome

Plants

Brazil nut	*Bertholletia excelsa*
Cacao	*Theobroma cacao*
Camphor tree	*Cinnamomum camphora*
Chicle	*Manilkara chicle*
False kola nut	*Garcinia kola*
Kapok tree	*Ceiba pentandra*
Kola nut	*g. Cola*
Makore	*Mimusops heckelii*
Monkey pot tree	*Lecythis ollaria*
Mopane	*Colophospermum mopane*
Oil palm	*Elaeis guineensis and Elaeis oleifera*
Trumpet creeper	*Campsis radicans*

Mammals

African golden cat	*Profelis aurata*
African brush-tailed porcupine	*Atherurus africanus*
African forest elephant	*Loxodonta cyclotis*
African pygmy squirrel	*Myosciurus pumilio*
Asian elephant	*Elephas maximus*
Asian golden cat	*Catopuma temminckii*
Asiatic brush-tailed porcupine	*Atherurus macrourus*
Aye aye	*Daubentonia madagascariensis*
Baird's tapir	*Tapirus bairdii*
Bearded pig	*Sus barbatus*
Binturong	*Arctictis binturong*
Black-shouldered opossum	*Caluromysiops irrupta*

(Continued)

Bongo	*Tragelaphus eurycerus*
Bonobo	*Pan paniscus*
Borneo orangutan	*Pongo pygmaeus*
Brazilian rabbit	*Sylvilagus brasiliensis*
Brazilian tapir	*Tapirus terrestris*
Brown-throated three-toed sloth	*Bradypus variegatus*
Bumblebee bat	*Craseonycteris thonglongyai*
Bush dog	*Speothos venaticus*
Bushy-tailed opossum	*Glironia venusta*
Cacomistle	*Bassariscus sumichrasti*
Chimpanzee	*Pan troglodytes*
Clouded leopard	*Neofelis nebulosa*
Collared peccary	*Pecari tajacu*
Crab-eating raccoon	*Procyon cancrivorus*
Crested genet	*Genetta cristata*
Crested gibbon	*Hylobates concolor*
Crested porcupine	*Hystrix cristata*
Duck-billed platypus	*Ornithorhynchus anatinus*
Eastern gorilla	*Gorilla beringei*
Eastern long-beaked echidna	*Zaglossus bartoni*
Feather-tail glider	*Acrobates pygmaeus*
Fishing cat	*Prionailurus viverrinus*
Fossa	*Fossa fossana*
Four-eyed opossum	*Metachirus nudicaudatus*
Gambian mongoose	*Mungos gambianus*
Gaur	*Bos frontalis*
Ghost bat	*Macroderma gigas*
Giant anteater	*Myrmecophaga tridactyla*
Giant armadillo	*Priodontes maximus*
Giant forest hog	*Hylochoerus meinertzhageni*
Giant otter	*Pteronura brasiliensis*
Giant pangolin	*Manis gigantea*
Golden Palace monkey	*Callicebus aureipalatii*
Greater naked-tailed armadillo	*Cabassous tatouay*
Gray brocket deer	*Mazama gouazoubira*
Gray gibbon	*Hylobates muelleri*
Grison	*Galictis vittata*
Ground pangolin	*Manis temminckii*
Hoffmann's two-toed sloth	*Choloepus hoffmanni*
Honey badger	*Mellivora capensis*
Indian pangolin	*Manis crassicaudata*
Indonesian tiger	*Panthera tigris corbetti*
Jaguar	*Panthera onca*
Jaguarundi	*Puma yaguarondi*
Javan rhinoceros	*Rhinoceros sondaicus*

Jentinks's duiker	*Cephalophus silvicultor*
Kinkajou	*Potos flavus*
Kloss's gibbon	*Hylobates klossii*
Leopard	*Panthera pardus*
Leopard cat	*Prionailurus bengalensis*
Long-tailed macaque	*Macaca fascicularis*
Long-tailed pangolin	*Manis tetradactyla*
Long-tailed porcupine	*Trichys fasciculata*
Malayan pangolin	*Manis javanica*
Malaysian tapir	*Tapirus indicus*
Maned three-toed sloth	*Bradypus torquatus*
Margay	*Leopardus wiedii*
Mountain tapir	*Tapirus pinchaque*
New Guinea bandicoot	*Microperoryctes papuensis*
New Guinea mouse bandicoot	*Microperoryctes murina*
Niger Delta pygmy hippo	*Choeropsis liberiensis*
Nile hippopotamus	*Hippopotamus amphibius*
Nine-banded armadillo	*Dasypus novemcinctus*
Northern naked-tailed armadillo	*Cabassous centralis*
Northern raccoon	*Procyon lotor*
Ocelot	*Leopardus pardalis*
Okapi	*Okapia johnstoni*
Olingo	*Bassaricyon gabbii*
Oncilla	*Leopardus tigrinus*
Pale-throated sloth	*Bradypus tridactylus*
Panther	*Puma concolor*
Pig-tailed langur	*Simias concolor*
Puma	*Puma concolor*
Pygmy hippopotamus	*Choeropsis liberiensis*
Red brocket deer	*Mazama americana*
Red giant flying squirrel	*Petaurista petaurista*
Ring-tailed lemur	*Lemur catta*
River otter	*Lontra longicaudis*
Sambar	*Rusa unicolor*
Short-beaked echidna	*Tachyglossus aculeatus*
Short-eared dog	*Atelocynus microtis*
Siamang	*Symphalangus syndactylus*
Silky anteater	*Cyclopes didactylus*
Slivery gibbon	*Hylobates moloch*
Sloth bear	*Melursus ursinus*
South American coati	*Nasua nasua*
Southern two-toed sloth	*Choloepus didactylus*
Sportive lemur	*g. Lepilemur*
Sugar glider	*Petaurus breviceps*

(*Continued*)

Sumatran rhinoceros	*Dicerorhinus sumatrensis*
Sumatran tiger	*Panthera tigris sumatrae*
Sun bear	*Helarctos malayanus*
Tamandua	*Tamandua tetradactyla*
Tayra	*Eira barbara*
Tree hyrax	*Dendrohyrax arboreus*
Tree pangolin	*Manis tricuspis*
Water civet	*Cynogale bennettii*
Water opossum	*Chironectes minimus*
Western gorilla	*Gorilla gorilla*
White-faced capuchin	*Cebus capucinus*
White-handed gibbon	*Hylobates lar*
White-lipped peccary	*Tayassu pecari*
White-nosed coati	*Nasua narica*
Yellow armadillo	*Euphractus sexcinctus*
Zebra duiker	*Cephalophus zebra*

Birds

African Gray Parrot	*Poicephalus senegalus*
Cassowary	*Casuarius unappendiculatus*
Crown Eagle	*Aquila heliaca*
Giant Blue Turaco	*Corythaeola cristata*
Harpy Eagle	*Harpia harpyja*
Henst Goshawk	*Accipiter henstii*
Monkey Eagle	*Pithecophaga jefferyi*
Oilbird	*Steatornis caripensis*
Quetzal	*Pharomachrus mocinno*
Rhinoceros Hornbill	*Buceros rhinoceros*
Shelley's Eagle Owl	*Bubo shelleyi*

Reptiles and Amphibians

Amethystine python	*Morelia amethystina*
Anaconda	*Eunectes murinus*
Basilisk lizard	*Basiliscus basiliscus*
Black caiman	*Melanosuchus niger*
Black iguana	*Ctenosaura similis*
Black mamba	*Dendroaspis polylepis*
Blanford lizard	*Acanthodactylus blanfordii*
Boomslang	*Dispholidus typus*
Burmese brown tortoise	*Manouria emys*
Burrowing python	*Calabaria reinhardtii*
Bushmaster	*Lachesis muta*
Carpet python	*Morelia spilota*

Emerald tree boa	*Corallus caninus*
Fer-de-lance	*Bothrops asper*
Five-banded lizard	*Draco quinquefasciatus*
Green mamba	*Dendroaspis angusticeps*
Green tree python	*Morelia viridis*
Home's hinged-back tortoises	*Kinixys homeana*
Indian cobra	*Naja naja*
Indian python	*Python molurus bivittatus*
King cobra	*Ophiophagus hannah*
Komodo dragon	*Varanus komodoensis*
Malaysian box tortoise	*Cuora amboinensis*
Nile monitor lizard	*Varanus niloticus*
Paradise tree snake	*Chrysopelea paradisi*
Reticulated python	*Python reticulatus*
Rock python	*Python sebae*
Royal python	*Python regius*
Serrated hinged-back tortoise	*Kinixys erosa*
Short-tailed monitor lizard	*Varanus brevicauda*
Spiny hill tortoise	*Heosemys spinosa*
Sulawesi lizard	*Draco spilonotus*
Surinam toad	*Pipa pipa*
Tokay gecko	*Gekko gecko*

Invertebrates

African giant swallowtail	*Papilio antimachus*
Blue morpho	*Morpho menelaus*
Kinabalu giant leech	*Mimobdella buettikoferi*
Nephila spider	*Nephila clavipes*
Owl butterfly	*g. Caligo*
Tiger leech	*Haemadipsa picta*

4

The Tropical Seasonal Forest Biome

The Tropical Seasonal Forest Biome, also called the Tropical Deciduous Forest, Monsoonal Forest, or Tropical Dry Forest Biome, is found in the tropics where a distinct seasonal climate is prevalent. The Tropical Seasonal Forest Biome once occupied a large area on all continents that lie along the Equator. These forests existed for millions of years and shifted in extent with changing tectonic and climatic events. Tropical seasonal forests are high in terrestrial biodiversity, second only to the tropical rainforests. Tropical moist and tropical dry deciduous forests, semi-evergreen seasonal forests, evergreen seasonal forests, and dry forests are part of this biome (see Figure 4.1). These forests undergo several months of severe or absolute drought. The different forests types are differentiated by water limitation, seasonality, and length of dry season, duration of leaves, vegetative structure, and substrate.

Plants and animals have adapted strategies to cope with this dry season. Many trees are deciduous, losing their leaves when the rains ceases. Other plants have developed small, hard, evergreen leaves that are drought resistant. Animals have also adapted: many will migrate into and out of the forest, others change their diet, and still others become less active.

Tropical seasonal forests are rich in biodiversity and historically have been the center of major human populations and agricultural development. The Tropical Seasonal Forest Biome has been reduced to a fraction of its original distribution, making it the most endangered of all terrestrial biomes. Despite its biological, geographical, and cultural importance, this biome is one of the least-known tropical ecosystems.

Figure 4.1 Seasonal forest at the beginning of the dry season in Guanacaste, Costa Rica. *(Photo by author.)*

Geographic Location

The Tropical Seasonal Forest Biome occurs globally in wide bands along the perimeter of the Tropical Rainforest toward the margins of the tropical latitudes between 10° and 20° N and S latitudes, at elevations below 3,000 ft (1,000 m). The biome occurs in three distinct geographic regions (see Figure 4.2). These regional

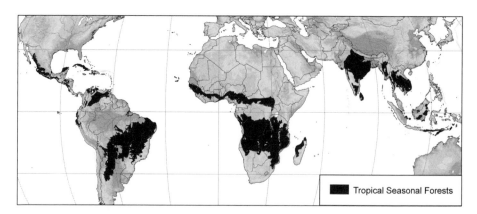

Figure 4.2 Distribution of Tropical Seasonal Forest Biomes around the world. *(Map by Bernd Kuennecke.)*

expressions have distinct characteristics and support unique species. Even within a region, seasonal forests can vary considerably from deciduous to evergreen dry forests.

In Mexico and Central America, the seasonal forests are found along the west coast as well as on several islands in the Caribbean. The South American countries of Brazil, Bolivia, Paraguay, and Argentina also hold substantial areas of Neotropical seasonal forests. Some of the most diverse dry forests in the world occur in southwestern Mexico, in the Chaco of South America, and the caatinga in northeast Brazil. Seasonal forests occur in Africa along the margins of the rainforests of West and Central Africa and are interspersed with savannas along the eastern coast. Madagascar has a remnant tropical dry forest along its western coast. In Madagascar, the seasonal forests support a unique assemblage of plant and animal species found nowhere else. The Asian-Pacific region—India, Southeast Asia, the islands of Indonesia, New Caledonia, and Australia—have seasonal forests greatly influenced by seasonal monsoons. These seasonal forests house an exceptionally diverse group of large terrestrial mammals. In each region, plants and animals have evolved similar adaptive responses to dramatic seasonal changes in rainfall.

Formation and Origin of the Tropical Seasonal Forest Biome

The origin of this biome is similar to that of the Tropical Rainforest Biome explained in Chapter 2 of this volume. Tropical seasonal forests developed along the margins of the rainforest, while the supercontinental landmass of Pangaea was present during the Permian. During the Mesozoic Era, Pangaea split into two large landmasses, Gondwana in the south and Laurasia in the north, creating large-scale climate change. The large Gondwanan landmass contained the tropical forests of South America, Africa, Madagascar, Asia, India, and Australia. As the tectonic plates began to separate, South America (with Antarctica) and Australia began to move away from Africa. In the early Cretaceous Period (around 120 mya), Madagascar and India separated from the southern landmass and traveled northeast, colliding with Asia. Madagascar became an isolated island in the Indian Ocean. Later, the Australian landmass broke off and traveled eastward. South America, Australia, and Africa became isolated island continents allowing for the evolution of a unique set of plants and animals. On the northern landmass of Laurasia, North America was connected to Europe and Asia and began to slowly move north.

Tropical seasonal forests were widespread around 65–50 mya, in the early Tertiary Period. Changing climate brought increasing seasonality, with a distinctive wet and dry season in parts of the tropics. This caused the tropical rainforest to shrink, and the seasonal forest to expand. Climate continued changing, shifting from seasonal to less seasonal throughout the Pleistocene. During interglacial periods, when warmer, wetter, and less seasonal patterns returned, the rainforests expanded and the seasonal forests became restricted and fragmented.

Mountain building and fluctuating sea levels significantly influenced the flora and fauna of the forests. Lower sea levels exposed land bridges from North to South America and through the Pacific Islands of the Sunda Shelf, allowing for the interchange of species and changing the composition of plants and animals within these areas. The return to higher sea levels left islands and continents temporarily isolated, providing the opportunity for species endemism.

Although the distribution of many taxa found in the tropical seasonal forests can be related to their past, the current structure and appearance of today's forest is probably very different. Long-term human occupancy in these seasonal forests, the use of fire, and the conversion of land for shifting agriculture have significantly changed the nature and extent of the world's Tropical Seasonal Forest Biome.

Climate

Temperatures in the Tropical Seasonal Forest Biome remain warm throughout the year. Average monthly temperature varies from 75°–81° F (24°–27° C), depending on the location of the forest. Temperatures show some seasonal variation, ranging from 68°–86° F (20°-30° C). Tropical seasonal forests remain frost free. The unifying climatic variable of all tropical seasonal forests is the strong seasonality of rainfall. Annual precipitation can be high, ranging from 40–80 in (1,000–2,000 mm), and occurs mostly in the summer months (see Figure 4.3). The Tropical Seasonal Forest Biome is characterized by a four- to seven-month dry season, sufficient to cause many trees, vines, and other plants to shed their leaves. The abundant seasonal rainfall, distinct dry season, and warm temperatures make this type of tropical climate a tropical wet and dry climate (Aw in the Koeppen classification system) or tropical monsoon climate (Am in the Koeppen classification), depending on the location.

Tropical seasonal forests vary considerably in the amount of rainfall and duration of dry periods. The amount and timing of rainfall is largely controlled by the seasonal shifts of the Intertropical Convergence Zone (ITCZ) and the occurrence of tropical monsoons. The ITCZ is a zone of low pressure, clouds, and rainfall that migrates north and south with the sun, creating a strong wet-dry seasonality in the tropical seasonal forest regions. Rainfall is concentrated in the periods when the ITCZ is present, with the dry season occurring when the ITCZ moves away. In the Northern Hemisphere, the dry period occurs from about December to March. Forest structure, canopy height, and total biomass are influenced by rainfall amounts and average length of dry season. See Chapter 1 for a full explanation of global circulation patterns and energy budgets in the tropics.

Monsoons
Several regions of the Tropical Seasonal Forest Biome, particularly Asia, are greatly influenced by monsoons. A monsoon is a significant shift of prevailing

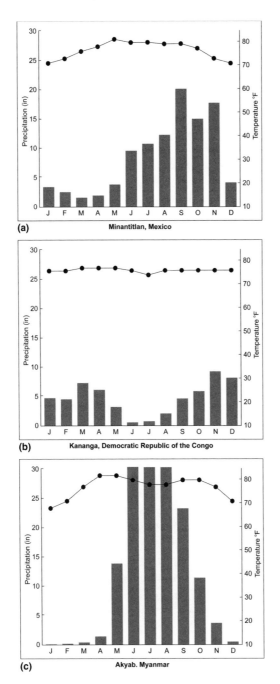

Figure 4.3 Climagraphs of Tropical Seasonal Forest Biomes. In each region, temperature remains high all year, but precipitation peaks in the summer and decreases in the
winter, creating a negative water balance. Myanmar is in the Southern Hemisphere,
with summer months from November to January. *(Illustration by Jeff Dixon.)*

winds caused by temperature differences between the land and sea and lasting for several months. Monsoons are seasonal shifts in wind direction, where moisture-laden air moves from ocean to land during the summer, and drier air moves from land to sea in the winter. Moist maritime tropical air moves from ocean to land, where it encounters hills and mountains that create uplift, cooling, and cloud development with intense convective precipitation. In the winter, the subtropical high-pressure system moves into more tropical latitudes and dominates with dry continental air and a prolonged dry season.

Although monsoons occur in a number of countries, the most well known of these is the Asian monsoon. The large Asian landmass, including Pakistan, India (and nearby Sri Lanka), Bangladesh, and Myanmar, and the large ocean surrounding the region (the Arabian Sea and the Indian Ocean) provide the perfect conditions for a monsoonal climate and a dramatic transfer of heat between land and sea.

Around April, premonsoon heat builds over the land, resulting in rising air and forming areas of low pressure over north India and the Himalayas. The ocean heats up more slowly, creating a temperature difference of as much as 36° F (20° C) between the land and sea. Over the oceans, the air is cooler and denser and linked to areas of high pressure. To maintain a balance of energy in the atmosphere, air begins to flow from the oceans to the land (high pressure to low pressure) bringing moisture-laden southwest winds across southern Asia (see Figure 4.4). With this shift, the rains come. Rains usually start in late May, hitting Sri Lanka and moving up from the Bay of Bengal into parts of northeast India and Bangladesh. As the land and ocean begin to cool in late summer and into autumn, the land loses heat more quickly than the ocean. The winds reverse and dry continental winds prevail, beginning the long dry season.

Monsoons also control much of the rainfall in the tropical seasonal forests of Africa. The high-intensity, high-water-content monsoonal airmasses can bring torrential rains over short periods of time. As much as 20–30 in (500–700 mm) of daily rainfall have been recorded in sites in Africa and India.

While the winter in the Northern Hemisphere experiences clear skies and a dry season, farther south, strong north and northeast winds originating from cold, northern Asia mix with moist tropical winds and bring severe weather including heavy rainfall and typhoons to Australasia. This climate pattern has created seasonal forests on several islands in the Pacific as well as Australia.

Soils

Soils of the Tropical Seasonal Forest Biome are varied, but tend to be similar to those of the rainforest (see Chapter 2). In South America and Africa, soils are ancient, with a high rate of leaching that washes away most of the nutrients. These older soils are derived from the Precambrian continental shield and are deeply

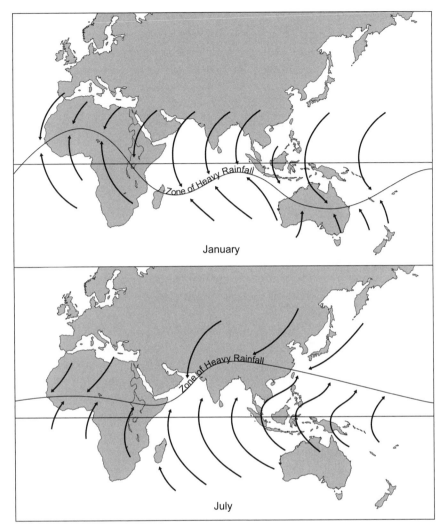

January

July

Figure 4.4 Diagram of Asian monsoon. Rains come in summer when warm marine area moves onto the land. *(Illustration by Jeff Dixon.)*

weathered, acidic, red and yellow oxisols. Oxisols are very low in nutrients and high in iron and aluminum oxides. Ultisols are also common throughout this biome and are the dominant soil in the tropics of Asia. These soils are derived from crystalline bedrock. They tend to be marginally fertile and less acidic. Additional soils found in tropical seasonal forests are well drained and fertile inceptisols and alfisols (mostly converted to agriculture) derived from basalt or limestone, and rich alluvial soils deposited from ancient and current rivers. Geology, topography, moisture, nutrient cycling, and decomposition rates all contribute to the composition and type of soils found in this biome. In general, soils of the tropical seasonal forests are less fertile and dominated by oxisols in the Neotropics, with more fertile soils of more recent origin occurring in the Asian-Pacific region. African soils lie somewhere in between the two regions. Specific soil types associated with large expanses of dry forest such as the Chaco and caatinga of South America are discussed in Chapter 5.

Nutrient Cycling and Decomposition

Tropical soils contribute little to the cycling of nutrients within the Tropical Seasonal Forest Biome. However, some soils provide needed phosphorus, magnesium, and nitrogen. The larger contributor of nutrients is the forest vegetation and organisms within the layer directly above the soil. Intense organic activity occurs in the decaying material dropped from plants and dead organisms. The decomposing layer rapidly increases at the start of the dry season when trees lose their leaves. This layer decomposes slowly until the wet season, when decomposing becomes rapid. The decomposition of dead plants and animals is undertaken by many organisms, including insects, aerobic and anaerobic bacteria, and fungi. They facilitate the conversion of unusable organic and inorganic compounds into useable nutrients available for plants.

The uptake of this decomposed matter is facilitated by roots and their accompanying beneficial fungi. The roots of many tropical seasonal forest trees are more extensive than their rainforest counterparts. Most of the root biomass occurs as fine roots found near the surface and at depths. They form networks with the fungi that rapidly absorb the nutrients and make them available to plants. The fine roots decrease during the dry season and grow again when the rain begins.

Tropical soils of the seasonal forest are complex and varied. Some are acidic, older soils with low fertility shaped by millions of years of constant heat and rainfall, others are infertile alluvial soils, and still others are fertile, younger soils created by years of sedimentation or volcanic activity. These richer soils have largely been used for agriculture, supporting the large human populations present in these regions.

Vegetation

Tropical seasonal forests vary by type and location. A diversity of forest structures, including variations in height, canopy layers, and density of trees are evident within the biome. Seasonal forests closer to the Equator are closed forests of semievergreen or largely deciduous trees. Other seasonal forests may be smaller and simpler in structure than these deciduous forests, with fewer canopy layers and more drought-resistant evergreen trees. Forest stratification, seasonal rainfall, and patchy distribution of soil moisture influence the type of forest and plant diversity in the Tropical Seasonal Forest Biome.

As the amount of annual rainfall in tropical areas fall below 78 in (2,000 mm), the number of woody deciduous plants increases. Deciduousness is a moisture-conserving response to water scarcity. In areas with a majority of months without rain, succulents and evergreen plants with smaller drought-resistant leaves start to dominate. Reduction in tree height is probably a result of reduced water availability at the root level, and increased duration of the dry season. Lianas and woody vines increase from wet to dry forests, accounting for up to 34 percent of all species in some tropical seasonal forests. The presence of epiphytes is reduced, likely the result of lower humidity and less dew during the dry season, creating an unfavorable environment. During the dry season, average humidity can be decreased to as low as 20–60 percent.

Tropical seasonal forests are high in overall biodiversity, though somewhat less species-rich than the tropical rainforest. In East Africa, for example, the coastal tropical seasonal forest is second to the rainforest in species richness. Researchers estimate that a lowland seasonal forest adjacent to a lowland rainforest can contain 50–100 percent of the total plant and animal species of the nearby rainforest. Farther from the rainforest, seasonal forests have, on average, 50 percent or fewer species than a comparable area of rainforest.

Forest Structure

Tropical seasonal forests represent a number of distinct communities, each with different structure and composition. Trees can be deciduous or evergreen with a canopy that varies from 10–130 ft (3–40 m) in height. The tropical seasonal forests that border the rainforest tend to have a closed or nearly closed canopy of deciduous trees. In taller deciduous forests, emergents can reach above the canopy to heights up to 145 ft (44 m). In the tropical deciduous forest, most of the canopy trees are deciduous and dormant during the dry season, although dormancy may not last the entire dry season. By dropping their leaves, trees are able to conserve water. Typically, leaf fall occurs when moisture stress is high at the start of the dry season. There may be little synchronicity in leaf fall within a given community, or even a single tree. Within an individual tree, some branches may lose their leaves while others sprout new ones. The leaves of deciduous canopy trees tend to be larger and less leathery than rainforest trees. They develop quickly at the start of the wet season. In other types of seasonal forests, evergreen trees dominate. These trees are small in stature and their leaves are often small and leathery to resist drying out during extensive periods of drought. Evergreen trees in the seasonal forest tend to be shorter, with an open or closed canopy. Most trees found in tropical seasonal forests tend to be shorter, with a less complex forest structure than the tropical rainforest. As the duration of the dry season increases, the canopy becomes open, with fewer layers, lower biomass, and lower net primary productivity compared with rainforests (see Figure 4.5). Tree diameters are smaller than those in the rainforest, and because of a short growing season, annual tree growth can be half that of the rainforest.

Families of trees found in the seasonal forest are often present in the rainforest; however, species are often quite distinct. Many trees within the legume family

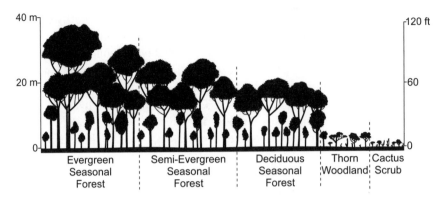

Figure 4.5 Transect from wet to dry forest. Trees become shorter and eventually disappear as rainfall decreases. *(Illustration by Jeff Dixon.)*

(Fabaceae) are common. Trees in the genus *Acacia* and *Caesalpinia* are found in all regions, with particular species found regionally. Trees in the fig family (Moraceae) are widespread in the seasonal forests around the world. Large trees in the kapok tree family (Bombacaceae), including the kapok and palo barrocho trees in the Neotropics and baobab in Africa and Madagascar, are among the tallest trees in these forests. Other common trees found within this biome include species from the melastome (Melastomataceae), oak (Fagaceae), euphorb (Euphorbiaceae), cashew (Anacardiaceae), laurel (Lauraceae), nutmeg (Myristicaceae), trumpet-vine (Bignoniaceae), mahogany (Meliaceae), and ebony (Ebenaceae) families. Palms are abundant in the canopy and understory in all regions. Dipterocarps are found in the seasonal forest of Africa and Asia. Teak (Verbena family) is a valuable hardwood found in the seasonal forests of Asia, particularly in Myanmar and Vietnam (see Chapter 5).

Deciduous trees in the tropical seasonal forest tend to have smooth, thin bark, while evergreen trees and those in the drier forests have thicker bark (see Figure 4.6). The thick bark may be an adaptation to fire. Some trees are armed with spines or thorns. Few trees have the large buttresses that are so common in the emergent trees of the tropical rainforest. Lianas and other woody vines often share dominance with canopy trees. Those found in the seasonal forest are less diverse than those in the rainforest. Epiphytes are typically decreased or absent in these forests, with the exception of some cacti, euphorbs, and bromeliads regionally.

As trees shed their leaves, the understory receives sunlight. In the deciduous forest, the understory shrub layer can be dense (see Figure 4.7). Many shrubs are evergreen or semi-evergreen, maintaining their leaves during the dry season to take advantage of the full sun available as the canopy opens. Some of these shrubs lose their leaves during the wet season when they become shaded.

Plant phenology, including the timing of leaf fall, flowering, and fruiting, is varied among different species and even individual trees of the same species.

Figure 4.6 Smooth bark on tropical trees prevents the growth of vines and can also inhibit animals from climbing the tree. *(Photo by author.)*

Although no conclusion has been reached as to the precise mechanism that signals plants to begin the process, water stress caused by decreased soil moisture is thought to be a significant influence. Day length and plant age are also important influences. Flowering typically occurs at the start of the dry season. Flowering and fruiting periods show patterns related to pollinators. Bird-pollinated plants tend to flower during the dry season, providing a food source for birds during a period of scarcity. The large and brightly colored flowers common on these plants are highly visible and attractive to potential bird pollinators during the mainly leafless dry season. The flamboyant tree, native to Madagascar has brilliant red and orange flowers that are produced at the start of the dry season when the tree starts to lose its leaves. Many of the plants that flower during the wet season are

Figure 4.7 Dense understory can be common in tropical deciduous forests where the canopy opens seasonally, as shown here in Guanacaste, Costa Rica. *(Photo by author.)*

pollinated by insects, or are located on drier sites, where moisture stress may inhibit flower initiation.

Seasonal forests are distinctive in the number of conspicuous flowers that have specialized pollinators such as hawk moths, bats, birds, and large- and medium-size bees. Although many of the plants in the seasonal forest are pollinated by animals, about one-third of all trees and 80 percent of lianas seeds are wind dispersed. In the rainforest, the majority of seeds are dispersed by animals.

Cauliflory is seen in trees of tropical seasonal forests. Cauliflory is the production of flowers and fruits on leafless trunks, rather than on twigs or smaller branches. Many trees in the seasonal forest flower (and later fruit) on very short leafless stems, or they develop flowers directly on the trunk or large branches. The flowers of cauliflorous trees are often pollinated by birds and bats. Their fruits are consumed by larger animals that may be unable to reach fruit in the canopy. These trees need these animals for the dispersal and germination of their seeds. Fig trees are an example of a common cauliflorous tree found throughout the Tropical Seasonal Forest Biome.

With increasing duration of the dry season and increasing latitude, seasonal forests move from deciduous to dry evergreen forests. With this change, the number of plants using a crassulacean acid metabolism (CAM) photosynthesis as a strategy for photosynthesis increases.

Underground biomass is higher in seasonal forests than rainforests. Roots are thought to be more extensive and deeper as an adaptation for storing and retrieving water and nutrients needed to flower during the dry season when photosynthesis is limited or nonoccurring. The development of a large rootmass allows for the storage of moisture and energy needed to carry out these functions. Some trees develop extensive horizontal roots, while others have deep roots. Many roots have an abundance of fine hairs. In a short dry forest, roots can make up as much as 50 percent of the total biomass of the forest. Many trees, lianas, and vines exhibit this adaptation.

Some tropical seasonal forests can appear almost lifeless during the dry season. Reddish-brown bark and gray branches dominate before the arrival of the rains. Leaves begin to appear, usually within 10 days following the first intense rain. Over the next few weeks, the leaves darken and turn the bare trees into a lush, dense forest with an abundance of plant and animal life.

Coping with Drought

Crassulacean acid metabolism (CAM) photosynthesis is an adaptation plants in arid regions have evolved to limit water loss while still carrying out photosynthesis. During the day, their stomata (plant pores) are closed to prevent the loss of water. In the evening, the stomata open and the plant absorbs carbon dioxide and stores it as malic acid. The final stage of photosynthesis is carried out during the day using the energy of the sun and converting the malic acid to carbohydrate. This process is common in desert plants and is named after the plant family Crassulaceae in which the process was first discovered.

Tropical Thorn Scrub or Thorn Forests

Tropical thorn scrub is found in all three regions in areas where precipitation is extremely low and seasonal. Annual rainfall can be as low as 20–25 in (500–640 mm), with the dry season lasting up to seven months. In Central America, this vegetation type is also called "cactus scrub" due to the abundance of cactus species present. Thorn scrub consists of low-growing, thorny trees, shrubs, and stem succulents. The sparse upper layer contains only a few species. The trees attain heights of 20–30 ft (6–9 m). The woody species are deciduous and small leafed; many have thorns or spines. The understory is poorly developed and not continuous, consisting of spiny and xerophytic shrubs and mosses, with a high proportion of dry land exposed at the surface. Acacias and other trees in the legume family are widespread throughout the thorn scrub forests of the world, cacti occur in the Neotropics, and members of the euphorb or spurge family are more common in Africa and Asia (see Figure 4.8). Stem succulents, xerophytic palms, epiphytic mosses and lichens, and—in the Neotropics—terrestrial bromeliads are common.

Tropical Woodlands

Tropical woodlands or savanna woodlands are areas of densely or loosely scattered trees with an understory of grass. Trees can be either evergreen with sclerophyllous leaves or deciduous. Tropical woodlands are susceptible to fire. Many of these woodlands have been replaced by savannas due to frequent fires, tree clearing for fuelwood, or grazing animals.

Specific forests or ecoregions within this biome include the caatinga of northeast Brazil; the Chaco of Paraguay, southern Bolivia, and northern Argentina; the

Figure 4.8 Thorn forests such as this one in Tanzania grow in sandy soils where a prolonged drought is common. *(Photo by author.)*

mopane and miombo woodlands of Africa; and deciduous dipterocarp forests of Asia. They are characterized by particular forest structure as well as species composition and are discussed in detail in Chapter 5.

Animals

Animal diversity in the Tropical Seasonal Forest Biome is similar to the rainforest, but this diversity decreases toward the subtropics. Studies of seasonal forest animals reveal an overwhelming diversity, second only to the tropical rainforest. Plant evolution and adaptation have led to diverse habitats and abundant opportunities for the diversification of animals. Myriad bird, reptile, amphibian, and mammal species have been recorded. Termites and other invertebrates also show high diversification. Many areas of the tropical seasonal forest have not been studied extensively, and many patterns of distribution as well as behaviors for seasonal forest animals have yet to be described.

Tropical Vertebrates

Mammal species are abundant in the Tropical Seasonal Forest Biome. Primates including marmosets, monkeys, and lesser and great apes can be found in these tropical forests. Large rodents, such as the pacas, and agoutis of the Neotropical seasonal forest provide food for animals and humans. Insectivorous animals, such as anteaters, and armadillos are confined to the Neotropics, while pangolins, aardvarks, some lemurs, and marsupial possums fill these niches in the forests of Africa and the Asian Pacific. Several mammals have developed complex stomachs or digestive processes to make efficient use of a limited diet of leaves. These include sloths of the Neotropics and langurs in Asia. Carnivores such as jaguars, panthers, leopards, and tigers, as well as weasels, mongooses, and civets are the main predators in the forest. Large herbivorous mammals such as elephants and rhinoceroses use the seasonal forests of Africa and Asia. Bats are a critical component of tropical ecosystems and are abundant in the tropical seasonal forests. Bats are important for plant pollination, seed dispersal, and insect predation.

Tropical seasonal forests share a high diversity of birds with the rainforest. Both seasonal and regional variations in abundance and diversity of bird families are evident. Different taxa have colonized and diversified in different geographic areas. Seasonal availability of flowers and fruit often brings migrating species of birds into the seasonal forests from nearby rainforest, savannas, and temperate forests of the middle latitudes.

Reptiles and amphibians are diverse but less abundant. Tropical snakes include venomous snakes, such as vipers, pit vipers, coral snakes, and cobras, as well as nonpoisonous snakes. Many are similar to those in the adjacent rainforest regions or bordering savannas. Many lizards, including skinks, iguanas, geckos, chameleons, and monitor lizards are found in these tropical forests. Frogs, toads, and caecilians are also present.

Adaptations. Although similar animal families and genera are found within the rainforests and seasonal forests, tropical seasonal forests contain a number of endemic species with unique adaptations that allow them to thrive under seasonal shifts in resources. These adaptations include local and regional migrations, changes in patterns of activity, changes in diet, and seasonal storage of fat or food resources.

Several vertebrates move seasonally to utilize various sources of food or shelter. For example, howler monkeys in the Neotropics will concentrate in the greener riparian forests during the dry season, and move back into the forest during the rainy season when leaves reappear. Hummingbirds move from riparian areas to dry forest during seasonal floral blooms, and bats move considerable distances to seek out trees bearing flowers and fruit. Similarly, elephant movement in Africa and Asia can be associated with seasonal availability of palatable food and water resources. Because many plants produce flowers, fruits, and seeds during the dry season, many animals move from adjacent habitats to take advantage of these food resources. Asian tigers will follow their prey as they move into the dry forest.

Changes in diet, activity patterns, and the timing of reproduction are additional adaptations many animals exhibit in response to seasonal water availability. During the dry season, some bat species change their diet from insects to more available, moisture-rich fruits. Anteaters will change their diet from ants to termites in the dry season, as termites have higher moisture content. Many amphibians are active for only a few months out of the year, with the period of activity coinciding with the warm, wet season. Several amphibians in tropical seasonal forests time their breeding to coincide with the wet season to take advantage of increased water availability.

Tropical Invertebrates

Most invertebrates found in seasonal forests are also found in the rainforest. Butterflies and moths, beetles, bees, roaches, mosquitoes, and stick and leaf insects are all abundant. Termites are plentiful year-round, playing a major role in the decomposition of organic matter and making nutrients available to vegetation. Termites are both terrestrial and arboreal in these forests and an important source of protein for many insectivores, especially in the dry season. Ants are extremely common in seasonal forests. Several species have evolved specialized mutualistic behaviors or affinities with certain plant species. By providing protection from herbivores, or competing plant species, ants find shelter and food in the trees. Beetles, mosquitoes, stick and leaf insects, katydids, leaf hoppers, and mantids have developed adaptive behavior, body structure, or coloration to adapt to their surroundings. Dragonflies are insect predators of the seasonal forests, hunting small insects including butterflies and mosquitoes. Scorpions, whip scorpions, and spiders call this biome home.

Distinctive climate and the evolution of a high diversity of plants and animals with specialized lifecycles and adaptive behaviors make the tropical seasonal forest a unique biome. Along with plants and animals, humans have adapted successfully to this biome, finding it to be well suited to their needs. People have been living in and using the products of these forests for tens if not hundreds of thousands of years. Long-term use has lead to the loss of much of the seasonal forests that once existed. Without a considerable effort to preserve, maintain, and sustain the tropical seasonal forests, this biome is in clear danger of disappearing.

Human Impact

Tropical seasonal forests have been used by humans for many years, and it is likely that any existing dry forest has been used as a source of firewood or charcoal production at one point in history. In Panama, for example, evidence of human modification of vegetation is more than 10,000 years old. Forest clearing and shifting cultivation has an even longer history in the Asian-Pacific region, one of the places where agriculture first developed. The anthropogenic uses of seasonal forest are so firmly placed in environmental history that the true nature and original extent of the seasonal forests may never be known. Many current savannas, scrublands, and thorn woodlands are believed to be the result of disturbances in dry forests.

The effects of widespread loss of tropical seasonal forests are wide ranging and profound. They include loss of biodiversity, climate change, changes in hydrological processes, erosion, soil compaction, and nutrient loss. Dry forests may be more vulnerable to conversion than other types of tropical ecosystems for various reasons. First, warm, dry climates tend to be preferable to hot, humid ones for both humans and livestock. Second, burning can be facilitated by the dry season, making the forest easier to clear than rainforests. Third, pest problems are reduced by a prolonged dry season, enhancing the desirability of dry forest land for agriculture. Insect-borne diseases such as malaria or dengue fever are less problematic in drier areas. Finally, dry soils are less prone to compaction than the soils of the rainforest, another desirable characteristic for agricultural land. In addition, dry forest soils are generally more fertile, less prone to nutrient leaching, and easier to manage with respect to successional vegetation and weeds. These advantages to settling in the seasonal forests have produced large population centers. As a result, the threats to this ecosystem are multiple and complex and deforestation rates exceed those in tropical rainforests.

Tropical seasonal forests are disappearing at an alarming rate and little undisturbed forest remains. As these forests are found in areas of the world with high populations and persistent poverty, complex social, political, and economic factors are involved in the process of deforestation. Shifting agriculture is one of the primary sources of seasonal forest loss, but spontaneous settlers and various government projects have also affected deforestation. Cattle ranching plays an important role in forest conversion in some regions, notably Central and South America. Road building into and through the forest can also have devastating effects—this is seen in the building of the InterAmerican and InterAmazonian highways through Central and South America.

Conversion of cropland for agriculture is considered one of the primary causes of deforestation in tropical seasonal forests because increasing populations lead to a dramatic rise in the number of shifting cultivators. Many of the issues discussed in the rainforest chapters occur in the tropical seasonal forests, but they occur at a faster rate due to increased population pressure. Sustainable agroforestry systems, although proven to be almost twice as productive as shifting cultivation, are not practiced often because many farmers lack the necessary technical knowledge to succeed at this alternative method.

The second primary cause, large-scale industrial agriculture, is on the rise and quickly becoming a major threat to survival of the remaining seasonal forests (see Figure 4.9). In the past, it was once considered no more of a threat than shifting agriculture, but that belief is quickly changing. Because land is often mistakenly considered to have little economic value until the forests have been cleared, the drier forest regions with fertile soil are being heavily targeted for conversion by large-scale agricultural operations. For example, in the seasonal forests of the southern Amazon, large expanses of land are being cleared for soybean farming. A similar situation is occurring near Santa Cruz in Bolivia, where deforestation rates are some of the highest in the world. The rush to convert land to soybean production is so intense that, despite selective logging of commercial species before

Figure 4.9 Large areas of land have been cleared for agriculture, leaving only small fragments of tropical seasonal forest remaining, as shown here in Guanacaste, Costa Rica. *(Photo by author.)*

completing clearing the forest, large amounts of valuable hardwoods are being burned. Production of this crop is considered one of the most important causes of deforestation in seasonal forests of the Neotropics.

Firewood removal and charcoal production are important staples in the lives of many people living in and around tropical seasonal forests and are responsible for a sizeable amount of forest degradation. Within the tropics, an estimated 70–80 percent of all harvested wood is used to meet the household needs of people in developing countries, and in Africa this proportion increases to 90 percent. Due to the growing demand for fuelwood and the slow growth of trees in the dry forests, a wood shortage is quickly occurring. Fuelwood is typically collected by family members for personal use; thus, the more families in an area, the more affected a forest will be. Two patterns of fuelwood collection are noted. The forested areas surrounding a population center are first depleted of wood and families must then travel farther to find wood. Fuelwood is more accessible in areas where logging operations have passed due to a supply of wood scraps left behind and the creation of roads further improving access to the forest. Once wood is no longer readily available, destruction of open forests and dry woody vegetation follows quickly. Extensive fuelwood collection in marginal areas can lead to total destruction of the forest and conversion to savanna or desert.

Given that the world's population is predicted to continue to increase, it seems unlikely that firewood needs will be able to meet the growing demand. One option to help alleviate the demand is the creation of plantation forests that can grow tree species with higher wood production yields. While promising, this alternative is limited by water and irrigation available to enhance growth. Irrigation can increase growth up to four times more than in naturally rain-fed plantations, but it is frequently too expensive to be feasible. Having to pay for fuelwood from plantations would be incredibly difficult for families that still rely on subsistence agriculture for survival.

Fire is a natural but relatively rare occurrence in most tropical seasonal forest types. The most vulnerable dry forests are those adjacent to savanna vegetation, although even those are not affected too negatively given the sparseness of vegetation under the tree canopy. If fire is frequent and seeds and seedlings are destroyed, however, the forests are unable to regenerate, and over time will become savannas or scrublands. Today, fire is commonly used by people to clear forests, control weeds in pastures, and burn logging debris. Both intentional and unintentional fires are growing in size and frequency throughout the tropics. Logged seasonal forests are more prone to fires, especially during droughts. Exposed forests dry out quickly and piles of woody debris are easily ignitable by lightning strikes, as well as by ranchers and farmers that practice shifting agriculture, who start fires to clear the forest floor. Fragmented forests are also more predisposed to fire because edges become desiccated and forest patches are often adjacent to fire-prone pastures and farmlands. Low-intensity fires, able to penetrate large distances into fragmented forests, can kill numerous trees and create canopy openings that further add to the forest's susceptibility to future catastrophic wildfires.

Small changes in the structure of a forest's ecological conditions, in combination with logging, slash-and-burn activities, and drought can increase the potential for fire. Droughts occurring during El Niño events have brought huge wildfires that swept through forests, destroying vast areas, particularly in Asia.

Finally, deforestation is not a random process, and not all forest types are equally threatened. Regions most vulnerable are those that are accessible with gentle slopes and productive, well-drained soils suitable for farming or ranching, such as the seasonal forests in Asia. Thus, one of the regions of greatest concern is Southeast Asia, because its tropical forests are smaller in area and are exposed to the highest relative rates of forest clearing and logging.

Tropical seasonal forests are the most threatened biome in the world. More than 80 percent of these forests are destroyed or degraded in Central America and Madagascar, while at least 60–80 percent has been converted in Africa and Southeast Asia. The two most contiguous areas of seasonal forest left are found in South America, but they too are at risk. In most other areas and regions, they are severely fragmented and scattered over extensive areas.

A great deal is left to learn about the ecology of tropical seasonal forests and its inhabitants. Increasing interest in the biology, function, and value of tropical seasonal forests gives hope that some of these unique ecosystems may be preserved. Sustainable land use policies based on an understanding of their ecology are urgently needed.

In this chapter, an overview of the Tropical Seasonal Forest Biome was presented. Each concept—climate, soil, vegetation, animals, and adaptations—is treated more specifically in Chapter 5 on the regional expressions of the biome.

Further Readings

Books

Bullock, S. H., H. A. Mooney, and E. Medina, eds. 1995. *Seasonally Dry Tropical Forests.* Cambridge: Cambridge University Press.

Janzen, H. D. 1988. "Tropical Dry Forest: The Most Endangered Major Tropical Ecosystem." In *Biodiversity*, ed. E. O. Wilson and F. M. Peter, 130–137. Washington, DC: National Academy Press.

Miles, L., A. C. Newton, R. S. DeFries, C. Ravilious, I. May, S. Blyth, V. Kapos, and J. E. Gordon. 2006. "A Global Overview of the Conservation Status of Tropical Dry Forests." *Journal of Biogeography* 33: 491–505.

Murphy, P. G., and Ariel E. Lugo. 1986. "Ecology of Tropical Dry Forest." *Annual Review of Ecology and Systematics* 17: 67–88.

Internet Source

National Geographic and World Wildlife Fund. 2001. "Wild World, Terrestrial Ecoregions of the World." http://www.nationalgeographic.com/wildworld/terrestrial.html.

5

Regional Expressions of Tropical Seasonal Forests

Several types of seasonal forests occur in the three regional expressions of the Tropical Seasonal Forest Biome in the Neotropics, Africa and Madagascar, and the Asian Pacific. Tropical seasonal forests are accommodating environments for humans, and large populations have settled into these regions. The role of humans in changing the tropical seasonal forest is profound. The clearing and conversion of tropical seasonal forests throughout the world make it the most threatened biome. This chapter describes the regional expressions of the Tropical Seasonal Forest Biome. Table 5.1 provides a quick comparison of the tropical seasonal forest regions of the world.

Neotropical Seasonal Forests

Tropical seasonal forests in the Neotropics occur from Mexico to Argentina (see Figure 5.1) and from sea level to 4,000 ft (1,200 m) elevation. The Neotropical region can be divided into two main subregions: Mesoamerica and South America (see Figure 5.2). Tropical seasonal forests in Mesoamerica include areas on the Yucatan peninsula and in Central America along a narrow strip on the Pacific coast from southern Mexico to Costa Rica. In Mesoamerica, the majority of seasonal forests occur in the rainshadow of the mountains or on limestone soils derived from marine sediments. Tropical seasonal forests also can be found in limited and fragmented distributions along the Atlantic in Belize and Honduras, and on the Caribbean Islands of Cuba, Puerto Rico, the Dominican Republic, and the Lesser Antilles. The South American subregion includes the Caribbean coasts of

Table 5.1 Regional Comparisons of Tropical Seasonal Forests

	NEOTROPICS	AFRICA	SOUTHEAST ASIA
Geographic location	Mexico, Pacific side of Central America, Gran Chaco, and caatinga	Sudanian and Zambezian Regions	Southeast Asia, India
Annual rainfall	27–78 in. (700–2,000 mm)	27–60 in. (700–1,550 mm)	40–120 in. (1,000–3,000 m)
Soils	Oxisols, ultisols	Oxisols, ultisols	Ultisols, inceptisols
Distinctive floral features	Palo borracho, barriguda trees, bromeliads, cactus	Mopane and miombo woodlands, baobabs	Teak, deciduous diptocarps, mast fruiting of trees
Distinctive faunal features	High bird diversity, armadillos, anteaters, migratory birds	Large primate diversity, large ground-dwelling mammals, elephant shrews	Large primates, high bird diversity, monitor lizards

Colombia and Venezuela and in the interior of South America, the caatinga of northeast Brazil and the Gran Chaco of Paraguay, Bolivia, and Argentina where an extensive area of dry forest is located. Smaller areas of tropical seasonal forest can be found in dry valleys of the Andes in northern Bolivia and Peru.

Figure 5.1 During the wet season, the Tropical Deciduous Forest looks similar to the rainforest, as shown here in Palo Verde, Costa Rica. *(Photo by author.)*

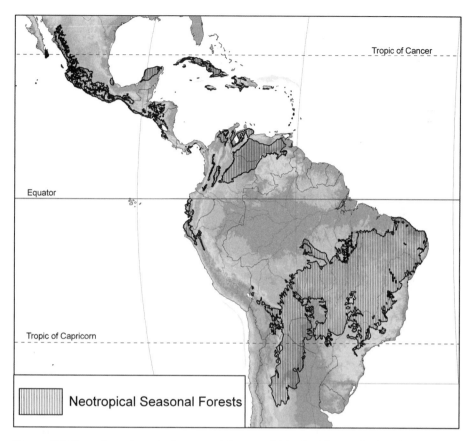

Figure 5.2 Location of tropical seasonal forests within the Neotropical region. *(Map by Bernd Kuennecke.)*

Origins of the Neotropical Seasonal Forest

Plate tectonics and fluctuating sea levels have greatly influenced the flora and fauna of Neotropical forests (see Chapter 3). In the late Permian Period, the continents were joined as the supercontinent Pangaea. At the break up of Pangaea, South America remained part of the southern protocontinent of Gondwana, along with Africa, India, Australia, and Antarctica. North America, Europe, and Asia were part of the northern protocontinent, Laurasia. At the breakup of Gondwana and the creation of the Mid-Atlantic Ridge around 150 mya, South America separated from Africa and remained isolated for millions of years. North and Central America did not separate from Europe until about 40 mya, when they moved toward South America. Moving plates and changing climate created fluctuating sea levels and periodic connections between South America and North America through land bridges. These connections dramatically changed the flora and fauna and lead to the extinction of many plant and animal species. The development of the Andes, about 15 mya, greatly influenced the South American subregion.

Climate

In the Neotropical seasonal forest, temperatures remain warm throughout the year. Average temperatures range from 73°–80° F (23°–26° C). Rainfall occurs during the summer and fall with total annual precipitation below 80 in (2,000 mm). These areas experience a short dry period (one to two months) during the summer and a longer dry period (two to six months) during the winter when little rain falls. Dry months vary in duration on the Atlantic and Pacific sides. Prevailing airmasses in the Neotropics flow from the oceans to the continents in both hemispheres. Seasonal forests are found on the leeward side and in the rainshadow of volcanoes and mountains, or in the interior of the continent. The shifting Intertropical Conversion Zone (ITCZ) plays a major role in the influx of rainfall in these seasonal forests (see Chapter 2). During the summer, the sun progresses toward the subtropics along with the ITCZ, and an increase in cloud cover and intense rainfall. In the winter, as the ITCZ moves away, a zone of subtropical high pressure with little precipitation develops. In Central America, tropical hurricanes can produce strong seasonal rainfall. Seasonal variations in moisture availability create a variety of seasonal forest types in the Neotropics.

During glacial and interglacial periods of the Pleistocene Epoch, tropical seasonal forests expanded and contracted. Some researchers suggest that during drier times, much of the Amazon rainforest reverted to seasonal forest and savanna, and when the Ice Ages ended, the rainforest expanded again, contracting and fragmenting the seasonal forests.

Soils

Soils are varied within the tropical seasonal forests of the Neotropical region. In most areas, soils tend to be very old and nutrient-poor. In other areas, soils are younger and fertile, a product of more recent volcanic activity. Underlying a large part of the South American subregion is one of the oldest rock formations in the world, the Precambrian Guyanan and Brazilian shields, and many of the soils in the region reflect that ancient history.

Soils within the Neotropics are classified into three main types: oxisols, ultisols, and inceptisols. Oxisols are deep red or yellow infertile soils, making up about 50 percent of all Neotropical soils. They occur primarily in areas influenced by the Precambrian basement rock. Ultisols are other weathered soils in the Neotropics. They have a high clay content and are slippery when wet and susceptible to erosion. Inceptisols occur on older alluvial plains along major rivers or are derived from volcanic activity and tend to be more fertile. Many of the forested areas where

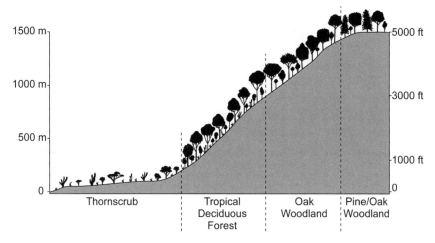

Figure 5.3 Gradient from coast to higher elevations in Mesoamerican Tropical Seasonal Forests. *(Illustration by Jeff Dixon. Adapted from Robichaux and Yetman 2000.)*

inceptisols dominate have been cleared and converted for agriculture. Well-drained fertile entisols and alfisols are extensively used for agricultural production.

Although seasonality of rainfall is the dominant influence on the biome, soil type can affect forest structure and composition. More drought-resistant evergreen vegetation is found on the more infertile soils, while lush deciduous vegetation is found on the fertile and deeper soils. Soil erosion is high when the forest is converted for other uses. Rates of soil formation are low, leading to little recovery of these tropical soils once they are disturbed.

Vegetation in the Neotropical Seasonal Forests

As in other regions, the Neotropical Seasonal Forest Biome includes several types of forest. The semi-evergreen and moist deciduous forests are found along the perimeter of lowland rainforests, and as latitude increases the forest becomes drier with dry evergreen forests dominating. In drier subtropical areas or at higher elevations, thorn forests and oak-pine forests start to emerge (see Figure 5.3)

Forest Structure

Seasonal forests are varied in the Neotropics, and most seasonal forests tend to be of smaller stature and simpler structure than the rainforest. However, the range in forest structure is considerable due to geography, soils, and duration and intensity of the dry season. In addition to seasonal resource availability, structure plays a role in creating a variety of microhabitats within the forests.

In the tropical moist deciduous forests, a complex four- to five-layered forest is present (see Figure 5.4). The top layer consists of widely spaced deciduous trees

Figure 5.4 Forest structure within a Neotropical seasonal forest. *(Illustration by Jeff Dixon.)*

50 ft (15 m) or more in height, often with one dominant species. In the next layer, various trees about 40 ft (12 m) tall co-dominate. Many are broadleaved with smooth, thin bark. Lianas, palms, and some cacti also reach the canopy layer. Woody vines and lianas can be abundant. Both trees and woody vines tend to have conspicuous flowers that appear at the start of the dry season. The majority of forest tree and vine species have seeds that are wind dispersed.

Beneath the canopy, tall shrubs and small trees up to 16 ft (5 m) in height form the tallest understory layer. Shrubs up to 3 ft (1 m) in height typically form the next layer. The lowest layer is composed of ground-level herbs that generally are present only during the summer rainy season. Epiphytes, with the exception of cacti and bromeliads, are typically sparse or absent in these forests.

In drier forests, the canopy is much lower with fewer layers. Dry evergreen forest trees in the upper canopy can be as tall as 23 ft (7 m), with a lower canopy at 10 ft (3 m). Woody vines increase as the forest gets drier. Depending on the forest, the understory can be dense or quite open, as in woodland areas. The ground layer is also varied, although there tends to be less vegetation on the forest floor.

Considerable underground biomass exists in seasonal forests because trees develop extensive root systems to store and retrieve water and nutrients during the dry season. Most trees and some vines have mycorrhizal relationships that facilitate nutrient and water absorption.

Many of the plant families in the seasonal forests are found in the rainforest, although the species are different. Diversity is about 50 percent of that in the Neotropical rainforest. Unexpectedly, the diversity within the dry forest region increases farther away from the wet forest. For example, the dry forests of Mexico tend to be the most diverse of the Neotropical seasonal forests, mostly due to their unique species composition. The legume (Fabaceae) family is by far the most species-rich family in all areas with the exception of the Caribbean where the myrtle (Myrtaceae) family predominates. Other common families found in Neotropical seasonal forests are the trumpet vine (Bignoniaceae), coffee (Rubiaceae), spurge (Euphorbiaceae), vochysia (Vochysiaceae), soapberry (Sapindaceae), and caper (Capparidaceae) families. Plants in the cactus (Cactaceae) and coca (Erythroxylaceae) families are common in the understory along with the plants in the bean caper (Zygophyllaceae) and wild cinnamon (Canellaceae) families. Groundcover is usually sparse in the closed canopy forest with bromeliads (Bromeliaceae), composites (Compositaceae), aroids (Araceae), mallows (Malvaceae), purslane (Portulaceae), and arrowroot (Marantaceae). Endemism can be high at the species level; however, no endemic families and few endemic genera are present in the Neotropical expression of the biome.

Mesoamerican Dry Forests

Mesoamerican seasonal forests include tropical deciduous, tropical semideciduous, and thorn forests. The legume family is the most species-rich family found in most of these forests. Euphorbs and myrtle families are important components of the forest flora. Herbaceous and woody climbing vines are also present. Species in the morning glory (*Ipomoea* spp.) squash, cactus, fig, legume, trumpet vine, and yam (Dioscoreaceae) families are present. The few epiphytes present in these forests include bromeliads, cactus, and orchids. The canopy is typically closed with little understory biomass. Fire is problematic in these forests as few trees tend to be fire adapted. Many have thin bark that is easily damaged, rather than the thick bark commonly found on trees that evolved with fire. Less than 2 percent of these forests are currently protected.

The dry forests of the Antilles tend to be short in stature and low in diversity. They have a high density of short- and medium-size trees with an even canopy and no emergent trees. The forests are typically found on limestone substrate or shallow and rocky soils along the coasts and experience high winds and frequent hurricanes. They are able to recover more rapidly from these disturbances than the forests on the mainland. The trees tend to be evergreen and sclerophyllous. Species in the myrtle family tend to dominant the islands. Species in the cactus, euphorb, caper, and bean caper families are also common. Since these forests occur on islands they are vulnerable to human exploitation and degradation. The introduction of nonnative species to the islands threatens the forests' survival.

The Gran Chaco

The Gran Chaco is the largest expanse of dry forest in South America. It lies between 17° and 33° S latitude, occupying parts of northern Argentina, southeastern Bolivia, and 60 percent of Paraguay, as well as a small portion of southwest Brazil. The Chaco is a mosaic of vegetation, including dry forests and woodlands, thorn scrub, palm savannas, grasslands, marshes, and salt flats. Gallery forests colonize along rivers and streams. The tropical thorn forest is the dominant forest of the Chaco. Thorn forests are dense forests of opuntia cactus and mesquite trees with bayonet bromeliads and star cactus in the understory. This makes it a rather inhospitable place for large mammals (including researchers). The palo borracho or drunken tree (kapok family) is a large evergreen tree in the Chaco, with a swollen trunk used to store water for survival during the dry season (see Figure 5.5).

Figure 5.5 Palo borracho tree of the Chaco are also called drunken trees for their ability to hold large amounts of water within their trunk. *(Illustration by Jeff Dixon.)*

The Chaco lies upon the margin of the Brazilian Shield that subsided during the Mesozoic and early Cenozoic eras (248–50 mya). During that time, the area accumulated thousands of feet of marine and terrestrial sediments that have created soils dominated by sands, silts, and loess (wind-deposited silt and sand). Sediments from the Bermeyo and Pilcomayo River continue to enrich the area. Average temperatures range from 66°–75° F (19°–24° C) but can goes as high as 82° F (28° C) in the warmer months and as low as 54° F (12° C) in the colder months. Mean annual rainfall varies from 10–47 in (250–1,200 mm) with a dry season that can last two to seven months.

Composition and forest structure vary with the amount of annual precipitation and soil type. The dominant vegetation is open dry forest woodlands with caha, palo borracho, acacia, candelabra cactus, and endemic trees in the Anacardiaeceae family (*Schinopsis* spp.) dominant. Cacti and bromeliads are found in the understory.

The Gran Chaco supports a great diversity of plants and animals including several endemics. It is home to an estimated 500 bird species, 150 mammals, 120 reptiles, and 100 species of amphibians. The Chaco is also home to three endemic armadillos, eight endemic mice and rats, and one cavid rodent. The lesser mara is a large, endemic cavid rodent that resembles a cross between a rabbit and a rodent. Its head is more rodent-like with small ears, but its body has strong hind legs for jumping. The giant tuco-tuco is a medium-size fossorial rodent found in the Chaco; it is similar in appearance and lifestyle to the pocket gopher of North America, but it is larger in size (average length is 15 in [38 cm]). Rediscovered in 1974, the endemic Chacoan peccary is one of the larger terrestrial endemic animals of the Chaco. This peccary was originally described from fossils and was thought to be extinct. The Chaco is also home to the jaguar, puma, giant anteater, tapir, and the night monkey (the only nocturnal monkey in the Neotropics). The birds of the Chaco include a large flightless bird called the Greater Rhea, the Chaco Chacalaca, Guira Cuckoo, and Little Thornbird, among others. It is the winter home for many North and South American migrants.

Most of the Chaco has been destroyed, especially in Argentina. Causes of this destruction include clearing for large ranching and agricultural operations, and road building. Research is still needed to understand the plants, animals, and ecology of the Chaco. A few areas are under protection but not enough to maintain the integrity and biodiversity of the Gran Chaco.

The Caatinga

The caatinga is a large area of dry forest and scrubland in northeastern Brazil that provides habitat for a diverse flora and fauna. The climate is hot and dry with the dry season lasting 6–11 months. The average annual rainfall varies between 10 and 40 in (250–1,000 mm), with average temperatures of 75°–80° F (24°–27° C)

The caatinga is one of the richest dry forests in the world. Vegetation on the caatinga is extremely heterogeneous and diverse. Caatinga vegetation ranges from low scrublands, with shrubby vegetation up to 3 ft (0.9 m) tall in shallow sandy soils to tall forests with trees 80–100 ft (25–30 m) in height associated with fertile soils that are high in nitrates (see Figure 5.6). The barrigudas or belly trees (Bombacaeae) are tall trees with swollen trunks that tower over the other vegetation, similar to the palo borracho trees seen in the Gran Chaco. Although the flora of the caatinga is poorly known, studies have identified a diverse and distinctive set of species. The biodiversity of the caatinga is made up of at least 340 species of vascular plants, of which 30 percent are endemic. Legumes, euphorbs, and cacti make up the 22–63 ft (7–19 m) tall canopy. Distinctive and endemic species include several species of cactus, bromeliad, butterfly sage, and members of the trumpet vine family. Biodiversity of vertebrates is relatively high with at least 80 mammal, 200 bird, and 47 reptile and amphibian species. Many of these species are shared with the humid forest and cerrado that borders the caatinga. The caatinga houses 2 of the 10 most threatened birds in the world, the Indigo Macaw and the Little Blue Macaw.

Figure 5.6 The caatinga vegetation is scrubby with scattered trees. *(Photo courtesy of Antonio Carlos de Barros Correa, Federal University of Pernambuco, Brazil.)*

The caatinga has been inhabited for more than 10,000 years with low population density. Cattle ranching has been the main economic activity during the past 300 years with cattle roaming freely for most of that time. At least 50 percent of the caatinga has been either completely converted from its native vegetation or significantly modified. Centuries of overgrazing have resulted in large-scale modification of the region. Unsustainable timber extraction and extensive, uncontrolled fires have degraded the caatinga. More recently, conversion to cotton cultivation is leading to the nearly complete destruction of caatinga in some areas. Other crops grown in the area include rice, sugarcane, sisal, cocoa, corn, and beans. The caatinga is highly threatened with very little (less than 1 percent) protected in parks or reserves.

Animals of the Neotropical Seasonal Forest

The seasonal forest's structure, habitats, and abundance and diversity of plants, flowers, fruits, and seeds provide resources for a great diversity of animals, many on a seasonal basis. Most have developed adaptive strategies to live with the seasonal fluctuations in resources. Vertebrates can be exclusive to this biome or a subset of species from nearby rainforests or grasslands. Forests that are more isolated and those that have been highly fragmented are less species rich. Forests near large rivers or in moist habitats can support a larger and more diverse population. The fauna of the seasonal forests of the Neotropics today is the cumulative result of millions of years of geologic, climatic, and biological events.

Mammals

Second only to the rainforest, more mammals inhabit the tropical seasonal forest than any other biome. Seasonal forests in the Neotropics include a mixture of families that evolved in North and South America, and others with origins from an ancient African connection. Mammals in these forests include marsupials and placental mammals representing most terrestrial mammal orders and many families (see Table 5.2). These include opossums, anteaters, armadillos, sloths, monkeys, rodents, ungulates, carnivores, and myriad bat species.

The marsupials of the Neotropics are all opossums. The opossum family is restricted to the Americas, with one species in North America and about 70 species in Central and South America. Those present in the seasonal forest include common opossums, woolly opossums, black-shouldered opossums, bushy-tailed opossums, short-tailed opossums, mouse opossums, and four-eyed opossums.

Three-toed sloths can be found in seasonal forests, but their populations are in decline because of forest fragmentation. Insectivorous anteaters and armadillos are found in the forests of the Neotropics. All tend to be specialized feeders, eating mainly termites and ants. The giant anteater roams the forest floor, and the tamandua hunts for ants in the trees. Anteaters will switch food preference during the dry season, feasting on termites instead of ants, because of their higher moisture content. Many armadillo species inhabit these forests, including the larger hairy armadillo, the northern and southern naked-tailed armadillos, the nine-banded and six-banded armadillo, and three species of long-nosed armadillo. The Chaco is home to three endemic species of armadillo. Armadillos specialize in ants, termites, and other forest insects. They too will shift their diets to a majority of termites during the dry season.

Bats are found worldwide, and seasonal forests in the Neotropics contain a great diversity. Bats are the most numerous mammals in the seasonal forest. All Neotropical bats belong to the suborder Microchiroptera. These bats use echolocation to locate prey. Common bat families present in the seasonal forest include sheath-tailed bats, leaf-chinned bats, mustached and naked-backed bats, fruit bats, brown bats, vampire bats, and free-tailed bats. Bats are frugivores (fruit eaters), nectarivores (nectar and pollen eaters), carnivores (meat eaters), insectivores (insect eaters), sangivores (blood eaters), and omnivores (generalists). In the seasonal forests, bats are crucial in regulating insect populations and pollinating flowers.

Primates are less abundant in the seasonal forests than the rainforests, as their primary diet consists of leaves and fruit, which are available only seasonally. Black, red, and mantled howler monkeys and capuchin monkeys are most common

Table 5.2 Mammals found in Neotropical Seasonal Forests

ORDER	FAMILY	COMMON NAMES
MARSUPIALS (METATHERIA)		
Didelphimorpha	Didelphidae	Opossums
PLACENTALS (EUTHERIA)		
Artiodactyla	Cervidae	Deer
	Tayassuidae	Peccaries
Carnivora	Canidae	Gray fox, coyote
	Felidae	Pumas, ocelots, and small cats
	Mustelidae	Weasels, grisons, skunks
	Procyonidae	Coatis, raccoons, ringtails
	Ursidae	Black bear
Cingulata	Dasypodidae	Armadillos
Chiroptera	Desmodontidae	Vampire bats
	Emballonuridae	Sheath-tail bats
	Megadermatidae	False vampire bats
	Molossidae	Free-tailed bats
	Mormoopidae	Ghost-faced bats
	Natalidae	Funnel-eared bats
	Noctilionidae	Bulldog bats
	Phyllostomidae	New World leaf-nosed bats
	Thyropteridae	Disc-winged bats
	Vespertilionidae	Evening bats
Lagomorpha	Leporidae	Rabbits
Pilosa (edentates)	Myrmecophagidae	Anteaters
	Bradypodidae	Three-toed sloths
	Megalonychidae	Two-toed sloths
Primates	Aotidae	Night monkeys
	Atelidae	Howler monkeys
	Cebidae	Capuchin monkeys
	Pitheciidae	Titis, sakis, uakaris
Rodentia	Agoutidae	Pacas
	Caviidae	Cavies
	Cricetidae	Packrats, brush mice
	Dasyproctidae	Agoutis
	Echimyidae	Spiny rats
	Erethizontidae	New World porcupines
	Heteromyidae	Pocket mice
	Hydrochaeridae	Capybara
	Geomyidae	Pocket gophers
	Muridae	Old World rats and mice
	Sciuridae	Squirrels
Soricomorpha	Soricidae	Shrews

Figure 5.7 White-faced capuchin monkeys are common within the seasonal forests of Central America, as shown here in Costa Rica. *(Photo courtesy of Thomas Phillips Jr., M.D.)*

within the region (see Figure 5.7). Squirrel monkeys are also present, but populations in the dry forest are declining. Black-tailed silver marmosets and night monkeys inhabit the seasonal forests of South America. Monkeys in the seasonal forests will often retreat to nearby riparian areas during the height of the dry season when fruits and leaves are no longer plentiful.

Tapirs are the only odd-toed ungulates (order Perissodactyla) in the Neotropics. With continued hunting and the rapid destruction of these forests, their populations are in steep decline. The even-toed ungulates (order Artiodactyla) are represented by two families: peccaries and deer. Peccaries are medium-size pig-like animals. They are mostly diurnal and feed on fruit, nuts, leaves, snails, and other small animals. Only three species of peccary exist in the world and all are found in the seasonal forests of the Neotropics. The collared peccary and the white-lipped peccary are distributed in Central and South America into Argentina (see Figure 5.8). The third species is the rare Chacoan peccary, which lives in the Gran Chaco in Bolivia and in western Paraguay. This animal was thought to be extinct but was rediscovered in the 1970s. Gray or brown brocket deer and white-tailed deer are present in the dry forests. They are browsers, feeding on new leaves, twigs,

Figure 5.8 Collared peccaries travel in small and large groups in the forest, as shown here in Guanacaste, Costa Rica. *(Photo by author.)*

and flowers in the dry forests. The brown or gray brocket deer is a small solitary deer restricted to South America. White-tailed deer are common in North and Central America and south into Bolivia.

Rodents are abundant in Neotropical forests. Based on geographic location, squirrels are highly variable in coat color and pattern. All squirrels in these seasonal forests are diurnal and arboreal. They feed on fruits, flowers, nuts, bark, fungi, and some insects (see Plate XV). Other rodents such as pocket gophers and spiny pocket mice are found in the dry forests of Mexico and Central America. Tuco-tucos are found in South America. Mexican deer mice, rice rats, white-footed mice, pygmy mice, cane mice, and climbing rats that nest in trees can be found in parts of the Neotropical seasonal forests. They occupy a variety of niches, feeding on fruits, plants, fungi, and invertebrates. Some mice and rats will cache their food, allowing them to survive in the forest during the dry season. Prehensile-tailed porcupines and the cavy-like rodents, such as agoutis, pacas, and maras, can be seen in these dry forests. Maras are restricted to the South American dry forests, while agoutis can be found throughout the region.

Four of the five carnivore families are found the Neotropics. They include the dog, raccoon, mustelid, and cat families. Tropical carnivores tend to be omnivorous, feeding on insects, fruits, and leaves in addition to vertebrates. The South American fox and maned wolf are found in the southern subregion of the biome. Raccoons, coatis, and ringtails are medium-size carnivore found in the dry forests;

cacomistles (similar to ringtails) live in the rainforest and migrate to the deciduous forest when food is available. These carnivores eat beetles, spiders, scorpions, ants, termites, grubs, centipedes, eggs, and even land crabs and fruit when available. They occasionally will take small vertebrates such as mice, lizards, and frogs.

Long-tailed weasels, grisons, skunks, and hog-nosed skunks forage on the ground. They have dense fur (which is considered valuable) and an extremely powerful bite that enables them to kill prey larger than themselves. Long-tailed weasels are common in North America extending into Central and South America to Bolivia. The grison and tayra occur in Central and South America.

Small cats like the jaguarundi, margay, and ocelot, and the larger puma and jaguar live in this region or migrate to the forest when prey is plentiful. All occur from Central America throughout South America; pumas are found north into Mexico and the United States. Most species are nocturnal.

Birds

Species richness of birds is lower in the dry seasonal forests than the humid rainforest. The total number of bird species found in the seasonal forests of the Neotropics (635 species) is similar to the number of species found in single sites of lowland rainforest. Local endemism is high in seasonal forests, surpassing that of the rainforest with almost 90 percent of those restricted to these forests endemic. Many birds are found in only one area of the seasonal forest, and few are found throughout the entire region. As many as 300 species use the seasonal forest as their primary habitat year-round. In any single area, about 60–80 species use the forest as their primary habitat.

Many birds use the forest at the start of the dry season when flowers and fruit are abundant. Species from higher latitudes in North and South America migrate to these forests in winter. Many migrating birds, such as warblers, vireos, yellow throats, flycatchers, and titmice winter exclusively in the dry forests of western Mexico. At least 109 species of migrating birds have been identified in the Mesoamerican subregion.

Puffbirds, motmots, and jacamars are large insectivorous birds endemic to the Neotropics with several species present in the seasonal forests. Smaller insect feeders endemic to the Neotropics are species of antbirds, antshrikes, flycatchers, gnatcatchers, woodpeckers, woodcreepers, wrens, and vireos. These birds exploit different layers including the ground, understory, and canopy, as well as different plants or parts of plants (bark, twigs, leaf undersides, and so on). Some species of antbirds will locate a traveling army ant colony and remain with them for days until the ants stop temporarily to reproduce. The bird leaves the colony each night and returns in the morning. Research is currently under way to discover how the antbird is able to relocate the constantly moving colony each day.

Hummingbirds are restricted to the Americas. They are brightly colored nectar-feeding birds of many sizes and diverse bill lengths. This allows each species to specialize on different flowers and nectars. They are found in the seasonal forest

Figure 5.9 Guans can be found in the trees near the forest edge, as shown here in Costa Rica. *(Photo by author.)*

during the wet season and at the start of the dry season when flowers are present. Parrots, parakeets, and the magnificent macaws are brightly colored fruit and seed eaters in the forest.

The Neotropical seasonal forest hosts a variety of ground birds, including terrestrial cuckoos, curassows, guans, chachalacas, and tinamous, as well as doves (see Figure 5.9). Their diets consist of small reptiles and insects, as well as fruit found on the forest floor.

The Neotropical dry forests house a variety of carnivorous birds. Diurnal raptors include hawks, hawk eagles, and a diversity of falcons and kites found along rivers and waterways. Owls are the nocturnal hunters within the forest. Vultures are important carrion feeders. In the dry forests of northwestern Costa Rica, the Black-headed Vulture is often seen perched in groups on rock outcrops or drifting high over the forest in search of food.

Forest trees provide shelter and nest-building material for many birds. They are also great places for hunting insects and other prey. In seasonal forests, Oropendulas build long-hanging nests. These nests help to keep snakes away from their young nestlings. Weaver birds in the African dry forests build similar nests.

Neotropical birds come in many shapes, sizes, and colors, with varied behaviors and feeding specializations. This diversity of appearance and lifestyle allows for a great variety of species to exist within the seasonal forests. Some endemic species are restricted to small areas of seasonal forests, others are temporary visitors, and still others are distributed worldwide.

Reptiles and Amphibians

Snakes, turtles, lizards, caiman, frogs, and toads are among the many reptiles and amphibians that inhabit the Neotropical seasonal forests. In coping with the long dry season, some species become dormant in the dry season living in burrows underground. Venomous pit vipers and coral snakes, constrictors, and other nonvenomous snakes can be found in this region. Pit vipers, the most deadly snakes in the world, range from North America into Central and South America. Pit vipers have sensory depressions or pits between their nostrils and eyes that they use to sense warm-blooded prey, mostly small mammals and birds. They have sharp needle-like fangs that can deliver a lethal dose of poison and affect the blood tissue or nervous system. The fer-de-lance, one of the most notorious of pit vipers, is found in the seasonal forests as well as the rainforest. Other pit vipers include forest pit vipers, palm pit vipers (also called eyelash pit vipers), and hognose pit vipers. Rattlesnakes are found in the drier forest and scrublands of Mexico and Central America. Constrictors are a large group of nonvenomous snakes within the Neotropics forests; they are called boas in this region. Boas have long wide heads with pointed snouts and wide bodies. They capture their prey by attacking and biting it, coiling around it and tightening their grip until they suffocate the victim.

Other reptiles in the Neotropical seasonal forest include lizards, geckos, tortoises, and turtles. Iguanas are a common lizard throughout the Neotropics (see Figure 5.10) and one of the larger lizards to inhabit the tropics. They can be greenish-brown when young, but become darker brown as adults. Male iguanas change color with the season, becoming brighter during the dry season when mating begins. They are nonpoisonous, feeding on insects when young, and mostly fruits and leaves as adults. Iguanas are found in trees and on the forest floor. Anoles are another type of iguanid. They are great at climbing trees and can be found perched on tree branches at great heights. They also hunt on the forest floor. Anoles eat mostly insects. Crocodiles and caiman commonly can be seen along riverbanks of seasonal forest. Caimans can be found from southern Mexico through northern Argentina. American crocodiles are restricted to Central America and the Caribbean. Both of these reptiles are listed as endangered in this region.

Amphibians of the Neotropics belong to three orders: salamanders and newts, caecilians, and the largest group, frogs and toads. Amphibians are less plentiful in the seasonal forest because many require water to reproduce and typically lay eggs in a gelatinous sac in ponds or streams. However, several frog and toad species can be found. Members of the large Neotropical frog family include coquis, chirping frogs, horned frogs, and the Chacoan burrowing frog. These frogs inhabit trees and shrubs, as well as leaf litter on the forest floor. Some have direct developing eggs; that is, the tadpole stage is skipped and a fully formed frog emerges from the egg. Microhylid frogs occur in the Americas from the southern United States to

Figure 5.10 Green iguanas are found in trees and on the ground in the tropical deciduous forests of Central America, as shown here in Guanacaste, Costa Rica. *(Photo by author.)*

Argentina, as well as in other tropical parts of the world. They include both arboreal and burrowing frogs that can remain dormant through the dry season. Toads are also common in the seasonal forest.

Many vertebrates show physiological and behavioral adaptation to living in the seasonal forest. Birds and nectar-feeding bats are able to migrate seasonally, traveling hundreds of miles into and out of the forest. Other species may move to higher elevations and other habitats during the dry season. Several amphibians and reptiles become inactive during the dry season, burrowing into the forest soil. Shifts in diets have been recorded in several mammal and bird species. A few mammals, such as spiny pocket mice and agoutis, cache large amounts of food that they depend on during the dry season. Constraints on moisture and food availability have lead to differing survival strategies among vertebrates. By adapting these strategies, vertebrate populations in the seasonal forest remain rich and diverse.

Insects and Other Invertebrates

The Neotropics is home to innumerable insects and other invertebrates. They play a vital role in the maintenance of the tropical seasonal forest as pollinators and decomposers, and provide essential nutrition to myriad animals that inhabit the forest. Arthropod populations vary seasonally with the largest populations occurring in the early and mid-wet season, and the lowest population occurring during the dry season. Insects can be the most significant herbivores of dry forest vegetation. The actual diversity of insects within the seasonal forests of the Neotropics is unknown.

Ants play a crucial role in destroying and recycling organic material. Leaf-cutter ants are a subgroup of the fungus-growing ants present in the Neotropics (see Chapter 3). Army ants are swarming ants that are important predators of the rainforest. Army ants are social ants with a queen, soldiers, and worker classes. They form large colonies reaching as many as 1 million individuals and practice group predation in which the pack swarms the prey, killing and dismembering it to bring it back to their temporary nest. They are nomadic and hunt on the ground and in trees. They are constantly on the move only stopping at underground nests or in hollow logs during their reproductive cycles. Large bullet ants are common in the seasonal forests of the Neotropics. They can be as large as 1 in (25 mm) long. They live at the base of trees or in tree cavities.

Forest Guards

Several species of ants have developed symbiotic relations with Neotropical plants. Aztec ants have developed a mutualistic relationship with the cecropia tree, protecting the tree from other herbivores, and in return consume the glandular nodules rich in carbohydrates produced at the base of new growth. Acacia ants will tenaciously clean the areas around an acacia tree, their home, quickly eliminating any plants that sprout in the immediate area. Other ants will defend their host tree from herbivores. Still others will live within acacia trees but provide no benefit to the tree and simply exploit the tree's resources.

Termites are important factors in decomposing and recycling in seasonal forests and are an indispensable food resource during the dry season for several mammals, birds, and reptiles. Termites are social animals living in large groups, in tree cavities, stumps, or on the soil surface (see Figure 5.11). Chapter 3 provides a detailed look at termites.

Neotropical butterflies and moths are highly diverse. A few families, most genera, and practically all species living in the Neotropics are endemic to the region. Families of Neotropical butterflies include the brightly colored swallowtails, the whites, and the blues. The nymphs are another highly diverse and abundant group of butterflies. Less is known about moths. Larvae of tropical moths are often plant eaters, but some are leaf miners, stem borers, flower feeders, or fruit and seed consumers.

The Neotropics host a vast array of other insects. Numerous beetle species are present. Some are brightly colored, others nondescript. Some common forest beetles include long-horned beetles, dung and carrion beetles, harlequin beetles, fungus beetle, and wood-boring metallic beetles. Cockroaches are another common decomposer found in the forest. Mosquitoes are common in many of the seasonal forests during the wet season, but thankfully most disappear in the drier months. Some may carry diseases such as malaria, yellow fever, and dengue fever among others. Once the forest is cleared, disease-carrying mosquitoes can be a problem for the human populations that settle in the area.

Spiders, whip scorpions, scorpions, and centipedes are common in the Neotropical dry forests. Orb spiders build webs of strong silk, ant spiders impersonate

Figure 5.11 Termites nest in the trees in the tropical deciduous forest to avoid flooding on the ground during the rainy season. *(Photo by author.)*

ants, and social spiders communally build large webs to trap prey. Bromeliad spiders live within terrestrial and arboreal bromeliads and can be abundant in these forests. Scorpions are common in dry forests, their sting can be toxic and irritating but rarely fatal to large vertebrates. Whip scorpions are another type of arachnid found in seasonal forests. Centipedes are common nocturnal predators. They have powerful jaws that inflict a painful bite. They also use poison to subdue their prey.

Seasonal forests in the Neotropics are diverse and serve an essential ecological role in maintaining tropical biodiversity. However, this great diversity of species is in serious jeopardy as these forests continue to be degraded and destroyed. This destruction can lead to the extinction of hundreds of plants and animals restricted to this biome and other species that rely on these forests for part of the year.

Human Impact on the Neotropical Seasonal Forests

Remote sensing images taken during the past 20 years indicate that tropical deciduous forests have undergone rapid changes. Estimates of deforestation vary greatly depending on the area and method of detection. In Santa Cruz, Bolivia, and Rondonia, Brazil, for example, the estimated rate of deciduous forest destruction is among the highest in the world (approximately 6 percent a year). Nearly 80 percent of the seasonal forests in Central America are also gone. Some of the main causes of deforestation experienced in seasonal forests were discussed in Chapter 4. The Neotropical seasonal forests are being converted largely for ranching or agriculture. Small- and large-scale ranching operations, as well as plantations of soybean, sugar cane, and palms, are quickly replacing the seasonal forests. The expansion of cattle ranching has led to extensive deforestation of Mexican seasonal forests. Increased international demand for low-cost beef and agricultural subsidies have led to large-scale ranching operations in Costa Rica, Brazil, and Bolivia. Timber extractions and an increase in small-scale shifting agricultural also contribute to forest destruction. Increases in population have lead to urban expansion and the creation of roads into these forested areas, further fragmenting them.

Wildlife trade in South America is responsible for the loss of millions of animals each year. These animals are taken for their skins and meat or for live trade. Several animals such as jaguars and peccaries are heavily hunted. Many wild cats and foxes, as well as parakeets, macaws, and iguanas, are used in legal and illegal international wildlife trade.

Large-scale conservation efforts have focused on tropical rainforests with little attention on the loss of tropical seasonal forests. Most of the dry forests in Central America, Venezuela, and Bolivia have disappeared, while the forests of Mexico have been severely eroded. Large-scale efforts to protect the remaining seasonal forests are under way in Mexico, Costa Rica, and the Chaco. Land conservation is essential for species survival. Reforestation efforts using local species have been somewhat successful. Creating protected areas along with developing sustainable and profitable land use practices for local economies will be the key for conservation of this region's tropical seasonal forests.

African Tropical Seasonal Forests

The African expression of the Tropical Seasonal Forest Biome includes broadleaf deciduous forests, evergreen dry forests, and woodlands that generally occur in two subregions along a band between 6°–13° N latitude in the Northern Hemisphere and 5°–20° S latitude in the Southern Hemisphere, from sea level to 3,200 ft (1,000 m) (see Figure 5.12). The northern or Sudanian subregion occurs along the eastern border of the West African rainforest and the northern border of the rainforest of the Congo into East Africa. The Sudanian subregion is severely affected, with only small remnants of forest found interspersed with the savanna. The southern or Zambezian subregion is larger in extent and is found along the southern border of the rainforests of the Congo. In the Zambezian area, despite increasingly intense use by humans, large intact areas remain. Fragments of tropical dry forests also occur along the coast of East Africa, extending from the tip of Somalia along coastal Kenya, Tanzania, Malawi, and Mozambique into Zimbabwe. Fragments of dry forest exist on the Cape Verde Islands, the islands in the Gulf of Guinea, and on the western coast of Madagascar.

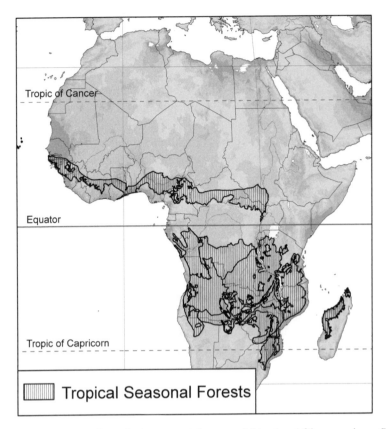

Figure 5.12 Location of tropical seasonal forests within the African region. *(Map by Bernd Kuennecke.)*

Origins of the African Seasonal Forests

The origin of the African seasonal forests is similar to the African rainforest discussed in Chapters 2 and 3. Precambrian basement rock underlies a large part of the area. Significant tectonic events and severe climatic episodes have influenced the current distribution of these forests. When rainforests reached their fullest extent between 35–10 mya, the seasonal forest was much smaller. During this time, the African continent became permanently connected to Asia (about 20 mya). This larger continental landmass created drier interior areas and a cooler, drier, and more seasonal climate in which the seasonal forests expanded and the rainforests became more restricted. Another cycle of global cooling with increased seasonality occurred at the beginning of the Pleistocene Epoch about 2.5 mya with the onset of alternating glacial and interglacial periods. This cycle of seasonal climate restricted the rainforest and expanded seasonal forests again. Glacial cycles continued throughout the Pleistocene with the last major glacial expansion ending about 18,000 years ago.

Climate

The common climatic factor in all tropical seasonal forests is the occurrence of a distinct dry period that can last from two to seven months. In African seasonal forests, rainfall during the dry season averages less than 4 in (100 mm) per month. During the rest of the year, rainfall is typically high, totaling 30–60 in (800–1,500 mm) or more. Rainfall totals and the duration of the dry season are the primary determinants of seasonal forest type. In some areas, rainfall can be greater, specifically at higher elevations on the windward side of mountains. The rainy season occurs in the summer and fall, with a cool dry season in winter followed by hot dry season in spring. Temperatures are warm, with an average minimum of 68°–72° F (20°–22° C) and an average maximum of 82°–92° F (28°–30° C). These areas do not experience frost.

The global circulation pattern that heavily influences the African Seasonal Forest Biome is the shifting Intertropical Convergence Zone (ITCZ). In Africa, a maritime airmass flowing southwest from the Indian Ocean meets an opposing hot, dry continental airmass from the northeastern deserts. Where the two meet is a zone of instability and high rainfall. These airmasses move seasonally from north to south with the shifting ITCZ, varying from 5°–7° N latitude in January to 17°–21° S in July. These movements account for the seasonal distribution of rainfall in tropical Africa. From December to March, during the dry season in the Northern Hemisphere, a hot dry wind known as the Harmattan Winds blow from the Sahara Desert, reaching West Africa. The Southern Hemisphere experiences warm dry winds from the Ethiopian Highlands.

Soils

Soils within the African region are varied. In some areas, soils tend to be very old and nutrient poor, originating from ancient bedrock of the Precambrian Shield. These oxisols are the dominant soils within the forests of the Zambezian subregion. Other soil types include ultisols and the more fertile alfisols and inceptisols. In general, soils of the seasonal forest are rather infertile and low in nutrients and minerals. This is due to the long-term processes driven by heat and moisture over hundreds of millions of years, as well as centuries of burning. Fires in nearby savannas have greatly influenced these forested landscapes and their soils. Although the seasonal forest is not conducive to fire, uncontrolled and frequent burning has changed many areas from dry forest to savanna.

Vegetation in the African Seasonal Forests

Tropical seasonal forests in Africa can be semi-evergreen or deciduous forests, evergreen dry forest, or woodlands having at least 70 percent continuous tree cover, and having more than three months of pronounced drought per year. Tree density and understory structure vary in each vegetation type. Seasonal forests in Africa cover 12 percent of the total land area of the continent, although these forested areas are rapidly decreasing. Seasonal forests are somewhat shorter, more open, and less species rich in the northern subregion than in the south. In the East African coastal forests, the forest is more susceptible to fire and easier to burn during the dry season. Burning provides usable land for cultivation and most of the seasonal forests in eastern Africa have been cleared for agriculture. The remaining patches are effectively "islands" within a matrix of savanna-woodland, coastal thicket, or farmland.

Forest Structure

The structure of African tropical seasonal forests differs from that of the Neotropics. The woody vegetation of the forests regions has three layers. In the semievergreen and deciduous forests, the upper layer is composed of tall trees up to 82 ft (25 m) that form a closed canopy. Lianas may grow as high as this layer. Beneath is a loose layer of trees reaching 30–50 ft (10–15 m) in height. The understory is composed of shrubs and vines that form a dense layer. Shade-tolerant grass species grow sparsely on the forest floor among the leaf litter. Typical species in the northern region are from the legume (Fabaceae), ebony (Ebenaceae), sopadilla (Sapotaceae), and soapberry (Sapindaceae) families. In the southern region, common tree species include the families above, as well as trees in the mahogany (Meliaceae) and cocoa/plum (Sterculiaceae) families. Several leguminous trees in the Caesalpinioideae subfamily dominate the woodlands, such as miombo trees in Central and South Africa.

Duration and intensity of rainfall and length of dry season vary throughout the African continent, with corresponding differences in leaf fall, flowering, and fruiting. For trees and shrubs, most leaf fall occurs at the start of the dry season and coincides with the main fruiting period. Individual trees do not all lose their leaves at one time, nor do all trees shed their leaves in unison.

Several distinctive forest types are found within this region, but they differ in species composition or dominance. The Sudanian subregion consists of dry woodlands with deciduous *Terminalia, Combretum, Isoberlina,* and acacia trees reaching 15–20 ft (507 m) with tall baobabs (reaching heights of 50 ft [15 m]) towering above them (see Figure 5.13). Like their palo borracho cousins in the Neotropics, baobabs can be enormous with swollen trunks.

The Zambezian *Cryptosepalum* dry forests are found in Zambia. These are evergreen forests confined to an area around the Kabompo River. These forests lie between the rainforest of the Congo and the Zambezian woodlands and make up the largest area of tropical evergreen forest outside the equatorial zone. The forest shares species from both rainforest and woodland-savanna and is species rich. The forests grow on infertile Kalahari sands with no permanent surface water. Because of this, these dry forests have remained relatively uninhabited. The forests

Figure 5.13 Structure of African dry forests. Trees tend to be shorter than those of the rainforest. *(Illustration by Jeff Dixon.)*

are dominated by *Cryptosepalum exfoliatum pseudotaxus* (Fabaceae), locally known as *mavunda*. The canopy is continuous at 50–60 ft (15–18 m) high, with a lower canopy at about 25 ft (8 m) and an understory shrub layer of bushes and lianas. The forest floor is often covered in mosses. Elephants, greater kudu, buffalo, duikers, and the red river hog use these forests. Bird diversity is high and includes cuckoos, drongos, flycatchers, bushshrikes, and weavers. Much of the forest is undisturbed due to its inaccessibility and infertile landscape.

Miombo Woodlands
Miombo woodlands are found throughout Sub-Saharan Africa and are located in southern Central Africa in the countries of Tanzania, Burundi, the Democratic Republic of the Congo (DRC), Angola, Zambia, and Malawi. Miombo woodlands are distinguished from other African savanna, woodlands, and forests by the dominance of tree species in the legume family, subfamily Caesalpinioideae, particularly miombo (*Brachystegia* spp.), mnondo (*Julbernardia* spp.), and *Isoberlinia* trees that tend to be restricted to this area. The canopy in these woodlands ranges in height from 15–33 ft (5–10 m), with little or no shrub layer but a grassy groundcover (see Figure 5.14). The grass layer can grow up to 6.5 ft (2 m) tall. The woodlands are interspersed with grassy plains and patches of denser forest. Most of the miombo trees and shrub species shed their leaves late in the dry season, and remain bare for a short time, usually less than three months. A few weeks to a month before the rains, the trees flush again with predominantly bright reddish or brilliant green new foliage. Most of the miombo trees and shrubs flower in this same period, immediately before the rains. Fire is an important ecological factor in miombo woodlands. The strong seasonality in precipitation leaves vegetation dry for several months

Figure 5.14 Miombo woodlands are a particular type of dry forest found in Eastern Africa, as shown here in Liwale, Tanzania. *(Photo courtesy of Roderick Paul Neumann, Ph.D., Florida International University.)*

and thunderstorms at the start of the rainy season can easily set the vegetation on fire. Miombo woodlands have been regularly set on fire by people for agriculture conversion, to improve pastures, or to flush out animals during a hunt.

Overall plant species diversity of the miombo woodland is high, though the faunal richness is moderate. Many of the plants and animals found in these woodlands are shared with nearby savannas or tropical rainforests. More than 170 different mammals are known to occur in these woodlands, including five endemic small species such as Vernay's climbing mouse and the miombo genet. Elephants and buffalo feed on the abundant but low-quality forage available. They are able to eat large quantities to make up for the lack of quality. Other large mammals such as zebra, sable antelope, roan antelope, and Lichtenstein's hartebeest are also present. Located on the northwest border of Tanzania, in the central miombo, chimpanzees are present in the Gombe Stream National Park, well known because of Jane Goodall's long-term studies here of the endangered chimpanzee. The area also supports red colobus monkeys, black-and-white colobus monkeys, blue monkeys, and red-tailed monkeys, as well as vervets and baboons. Ground pangolins and aardvarks feed on the numerous ants and termites found in the miombo.

Other large mammals found in miombo woodlands include lions, leopards, cheetahs, African wild dogs, jackals, and spotted hyenas. Smaller carnivores include servals, caracals, and the miombo genet. Bird diversity is high with few endemics. Typical miombo species include the Miombo Gray Tit, Miombo Rock-thrush, and Böhm's Flycatcher. The herpetofauna is only moderately rich, with a few endemic frogs and snakes, including the Angola ornate frog and the Huila forest tree frog and one endemic snake, Bocage's horned adder. Invertebrates (termites and caterpillars in particular) are important ecological agents in miombo woodlands and probably remove more biomass than large mammals. Termites are widespread and produce enormous mounds throughout the miombo region. These mounds change soil properties and produce patches rich in nutrients and organic matter within an otherwise nutrient-poor landscape.

In some areas like the Angolan miombo, low human populations and the movement of many rural people to cities with better security during times of civil war have left large stretches of habitat unaffected by human settlement. The floral biodiversity is thus relatively well preserved, although the animals remain at risk due to hunting. In the central Zambezian miombo woodlands, higher population densities have degraded most of the woodlands. Bushmeat hunting, dryland agriculture, deforestation from charcoal production near larger towns, and mining are increasing threats.

A number of protected areas contain miombo woodlands, but many of these have been effectively abandoned because of long civil wars. They have become open to poachers, human settlement, and cultivation. The impact of war on the fauna has been catastrophic throughout Angola and the Democratic Republic of Congo, with few if any viable populations of large mammal species left. Several areas in Tanzania and Zambia that contain miombo woodlands are better protected.

Mopane Woodlands

Mopane woodlands are widespread in low-lying areas of eastern Africa. Soils here were derived from Precambrian basement rocks as well as volcanic and sedimentary rocks. These woodlands are found primarily at low elevations on gentle slopes or the floors of river valleys. The dominant tree is mopane, which can form dense stands. Other trees include blackwood, leadwood, acacia, and baobab. The understory can be dense shrub or thick grasses depending on the soil and moisture content. Mopane provides food and shelter for African elephants that are frequently found in these woodlands and can be a major factor in shaping the vegetation. Fire is another factor. Cattle grazing and agriculture have degraded many of these woodlands. Few are in protective status. Government conservation areas, private game reserves, nature reserves, and conservancies within East African countries continue to provide some aspect of protection for the flora and fauna in this seasonal forest type.

Dry Forests of Madagascar

Dry deciduous forests are found along the western side of Madagascar, with a small remnant forest in the north, above 3,330 ft (1,000 m). These forests are some of the most unique forests in the world. They are high in endemic plants and animals at the species, genera, and family levels. They are found between the succulent scrublands of the southwest and the subhumid forest to the north and east. Trees in the legume, trumpet vine, euphorb, soapberry, sumac, and fig families are represented. Species from the milkweed family are the common lianas in the shrub layer. Distinctive plants in these forest include seven endemic species of Baobab tree (Africa has one species). Madagascar palm and blackwood are also present in the island's dry forests, along with flamboyant trees with bright orange and red flowers.

Several endemic animals are also found in the seasonal forests of Madagascar. The Angonoka tortoise is one of the world's 10 most endangered animals and is restricted to the dry forests of Madagascar. Unfortunately, due to long-term hunting of these animals, increased forest fragmentation, and fires, they are critically endangered with fewer than 1,000 individuals estimated to be left. These dry forests are also home to eight endemic lemurs. Lemurs are critical to the regeneration of the forests because they are important seed dispersers. Several endemic mice and rats including the forest mouse, western forest rat, and jumping rat (the size of a rabbit) inhabit these dry forests.

It is estimated that 97 percent of the Malagasy dry forest has been destroyed or severely altered due to fire and forest clearing for shifting agriculture and livestock pastures. Forest trees have been taken for firewood, charcoal production, and construction. Much of the remaining forest is susceptible to degradation due to its proximity to savannas where uncontrolled burning takes place. Increasing populations and encroachment into the dry forests of Madagascar continues to threaten the last remaining forest fragments. Parks and protected areas have been created with the cooperation of the Malagasy government, traditional leaders, and local organizations in an attempt to conserve the last remaining dry forests on the island.

Cape Verde Islands

The Cape Verde Islands are located in the mid-Atlantic about 300 mi (450 km) off the west coast of Senegal. They are of volcanic origin and support a small fragment of dry forest that still maintains a number of endemic species. On the islands, the dry season occurs between December and July, with a short wet season from August to November. The Cape Verde Islands are home to 15 endemic species of lizards, including giant skinks, giant geckos, mabuya skinks, hemidactylus lizards, and Torentola geckos.

Animals of the African Seasonal Forests

African seasonal forests have high rates of species richness and endemism. For example, the remaining eastern African coastal forest provides habitat for about 50 mammal species, 200 bird species, and 1,000 to 1,500 species of higher plants. These seasonal forests are quite diverse and share most of their species with the rainforest or savanna. Unlike the rainforest, dry forests (with the exception of

Madagascar) do not hold many endemic species. Because many of these areas remain inaccessible, less is known about the extent and ecology of the animals of this biome.

Mammals

Mammals in the seasonal forests of Africa include a mixture of families found in the savanna or rainforest. Many of these mammals spend part of their time in the seasonal forests when food is abundant and retreat to riparian areas or rainforests during the dry season. The southern seasonal forests share many mammal species with nearby savannas. Savanna mammals migrate in and out of the savanna into dry forests. Twenty-five different mammal families occur in these African forests (see Table 5.3).

Insectivorous pangolins, aardvarks, tenrecs, and shrews are found in the seasonal forests and scrublands in Africa. All tend to be specialized feeders, eating mainly termites and ants. Pangolins will switch food preference during the dry season, feasting on termites instead of ants, because of their higher moisture content. Pangolins are typically small, solitary, nocturnal mammals. Tenrecs are a diverse family of mammals that have evolved to filled many niches within the tropical forests of Madagascar. They are highly variable in form and size, ranging from small mouse to cat in size. Most tenrecs have pointed snouts and small eyes. Tenrecs have a limited distribution on the African continent, but a large and diverse group live on the island of Madagascar. Aardvarks are the only member of their order (Tublidentata) and are endemic to Africa. They look a bit like large long-snouted pigs with large ears. They can weigh 88–200 lbs (40–100 kg) and are excellent diggers. Aardvarks take advantage of the abundance of ants and termites found in the seasonal forests and hunt at night.

Bats live in most seasonal forests in Africa. They are abundant and both major groups of bats, Megachiroptera and Microchiroptera, are present. Megachiroptera, large fruit-eating bats, are often seasonal residents, taking advantage of the abundance of fruit at the start of the dry season. Microchiroptera are the small bats that use echolocation to navigate and locate prey. Most bats are nocturnal and rest during the day. Bats are important consumers of insects and fruits and are agents for seed dispersal and are pollinators for many tropical plants. Several bat species recently have been identified as a vector for the transmission of several deadly diseases in Africa.

Rodents are a very large and successful order in Africa. Squirrels, gerbils, spiny mice, pouched rats, and mole rats are residents of the seasonal forests. Many of these species are nocturnal, with the exception of squirrels and a few mice and rats. Several are important consumers of seedlings, seeds, and insects, and play a major role in forest dynamics and regeneration. Squirrels are the most abundant arboreal rodent in the forest and vary in size and color. The forest also houses a few flying squirrels that glide from tree to tree eating fruits, flowers, bark, and sometimes insects. The pygmy scaly tailed and the long-eared scaly tailed flying squirrels are

Table 5.3 Mammals Found in the Seasonal Forests of Africa and Madagascar

ORDER	FAMILY	COMMON NAMES
Afrosoricida	Tenrecidae	Tenrecs and otter shrews
Artiodactyla	Boviidae	Antelope
	Suidae	Pigs
Carnivores	Caniidae	Jackals
	Felidae	Leopards, lions, servals, caracals
	Herpestidae	Mongoose
	Hyaenidae	Hyenas
	Mustelidae	Ratels
	Viveridae	Civets and genets
Chiroptera	Emballonuridae	Sac-winged, sheath-tailed bats
	Hipposideridae	Leaf-nosed bats
	Megadermatidae	False vampire bats
	Molossidae	Free-tailed bats
	Myzopodidae	Old World sucker-footed bats
	Nycteridae	Slit-faced bats
	Pteropodidae	Old World fruit bats
	Rhinolophidae	Horseshoe bats
	Rhinopomatidae	Mouse-tailed bats
	Vespertilionidae	Insect-eating bats
Hyracoidea	Procaaaviidae	Hyraxes
Macroscelidea	Macroscelidae	Elephant shrews
Pholidota	Manidae	Pangolins
Primates	Cercopithecidae	Old World monkeys
	Cheirogaleidae	Dwarf lemurs
	Daubentoniidae	Aye ayes
	Galagidae	Galagos
	Hominidae	Chimpanzee
	Indriidae	Indris, sifakas
	Lemuridae	True lemurs
	Lepilemuridae	Sportive lemurs
	Lorisidae	Pottos
Proboscidea	Elephantidae	Elephants
Rodents	Anomaluridae	Flying squirrels
	Gliridae	Door mice
	Hystricidae	Old World porcupine
	Muridae	Old World rats and mice
	Nesomyidae	African and Malagasy endemic mice, rats
	Sciuridae	Squirrels
Soricomorpha	Scoricidae	Shrews
Tubulidentata	Orycteropodidae	Aardvarks

found in the seasonal forests. Pouched rats and mice, like the Gambian rat, are native to Africa and live in seasonal and scrub forests. Pouched rats can carry large quantities of food that they store and use to survive in the dry season when food is scarce.

Tree hyraxes are medium-size rodent-like animals with small ears, short legs, and no tails. They are found in the trees of the forest. Rock hyraxes are found in the drier scrubland areas where rocky outcrops are present (see Figure 5.15).

Primates are abundant in the seasonal forest, although not as diverse as those in the rainforest. Mona monkeys, vervets, and several species of colobus monkeys are seen in the trees, while mandrills and baboons are found searching for food in trees or on the ground. Chimpanzees are the only member of the great ape family present in the seasonal forests of Africa. Chimpanzees are primarily frugivores, but also consume seeds, nuts, flowers, leaves, pith, honey, insects, eggs, and vertebrates, including monkeys.

Many species of lemurs are found in Madagascar and nowhere else. Lemurs through adaptive radiation have evolved to fill niches taken by squirrels, rats, and monkeys, as well as some birds. Although many species have become extinct, eight species live in the dry forests of western Madagascar, including the mongoose lemur, golden-crowned sifaka, Perrier's sifaka, Milne-Edwards's sportive lemur, brown lemur, and three species of mouse lemurs.

African seasonal forests have a large number of ground-dwelling mammals, including those found in savannas and woodlands. The largest of these mammals is the African elephant, who finds shelter in the dry forests and a seasonal bounty of fruits (see Figure 5.16). Many other large African herbivores are present in the seasonal forest as permanent residents or migrants from nearby rainforests or savannas. These include zebra, bongos, roan antelope, sable antelope, greater kudu, common eland, wildebeest, blue and yellow backed duikers, impalas, buffalo, forest hogs, and bush pigs. The black rhinoceros is found in some Zambezian woodlands but is locally extinct in other areas.

Carnivores in the forest include hyenas, mongooses, civets and genets, African wild dogs, smaller cats, leopards, cheetahs, and lions. These are the main predators in the forest. Hyenas are more common in savannas, but spotted and striped hyenas also have been found in dry forests and scrublands. Genets, civets, and

An Unexpected Hero

The Gambian pouched rat is the largest murid rodent in the world. This nocturnal rodent can grow as large as a raccoon. They are often hunted for food. Gambian rats have the potential to save lives. Gambian rats are being trained to recognize explosives (through smell) and detect land mines in Africa. The rat is conditioned to associate the smell of TNT with a reward of food, training the rat to seek out the explosive. These rats are well suited for the job. They are adapted to the hot African climate and have an excellent sense of smell. Gambian pouched rats are smaller and lighter than dogs (typically used in the detection process), which severely decreases their chance of triggering the mine. Gambian rats are much cheaper to obtain and train than their canine companions. Once the rat has discovered the buried mine, it will scratch at the soil, and the mine will be identified and destroyed.

Figure 5.15 Rock hyraxes are found on rocky outcrops in the drier forest and savannas of Africa. *(Photo by author.)*

mongooses are medium-size carnivores that are mainly terrestrial, eating insects and small vertebrates. The fossa is a unique species in this family that occurs only in Madagascar. It is cat like in appearance and is the largest carnivore on the island, with its main habitat being the dry forests. It hunts on the ground and in the trees, taking birds, eggs, lemurs, rodents, and invertebrates.

The African golden cat is rare in seasonal forests of Africa. Other small cats, such as servals and caracals, are more abundant. Leopards are solitary, nocturnal large cats in the forest; cheetahs and lions hunt prey in the open dry forests and woodlands.

Birds
A variety of bird species live within the African seasonal forests region. Some are permanent residents, while others migrate from the rainforest or savanna to the dry forest when food is plentiful. Still others migrate from northern and southern mid-latitudes to spend the winter in these forests. The most common birds seen are cuckoos, tinkerbirds, greenbuls, woodpeckers, cuckoo shrikes, bush shrikes, flycatchers, drongos, waxbills, hornbills, and the endemic turacos. Several species of brightly colored sunbirds are present along with weaver birds with their long pendulous nests in the forest trees. Guinea fowl, spurfowl, and francolins roam the forest floor. Predatory and scavenging birds are also found in the seasonal forests. Many species of

Figure 5.16 African elephants of the dry forests are larger than those of the forest, as shown here in Samburu National Park, Kenya. *(Photo by author.)*

kites, goshawks, eagles, hawk eagles, snake eagles, buzzards, and vultures roam the skies searching for prey in the dry forests and scrublands. The Secretary Bird is found in the more open woodlands of this biome. Birds play an important role in the African Rainforest; they consume fruits and disperse seeds, eat invertebrates, reptiles, frogs, rodents, primates, and other birds. Healthy populations of many of these tropical birds rely on intact forests that are quickly disappearing.

Reptiles and Amphibians
The warm climate of the African tropics makes it an ideal home for cold-blooded reptiles and amphibians. A variety of lizards, turtles, chameleons, and snakes, as well as a few frogs and toads are found in the varied seasonal forests. Research is still needed in identifying and understanding the ecology of the herpetofauna of African seasonal forests, as intensive research in the forest is often difficult because of accessibility.

More than a hundred different species of snakes are found in the tropics of Africa. The rock python, one of the world's largest snakes, was once abundant in the seasonal forests, but due to heavy human predation and loss of habitat, they are restricted mostly to reserves and savannas. Many dangerous snakes inhabit the

African seasonal forests, including the golden vipers, rhinoceros vipers, bush vipers, mole vipers, night adders, puff adders, snake eaters, and quill-snouted snakes. Elapids such as forest cobras, black spitting cobras, and green and black mambas reside in the seasonal forests. Mambas are considered the most dangerous snakes in Africa. The black mamba is the largest venomous snake in Africa. Its extremely potent venom attacks the nervous system and is 100 percent fatal without antivenom. Mambas have been reported to be able to bring down animals as large as buffalos with their venom.

Lizards are perhaps the most common reptile in the African seasonal forest. Monitor lizards, agamas, skinks, and chameleons inhabit the forest. Lizards' primary food is invertebrates, but some will eat soft leaves from young plants. Monitor lizards are the largest lizard in Africa—the Nile monitor lizard can grow to 6.5 ft (2 m) long. Agamas, also known as Old World iguanas, are widespread throughout Africa and Asia. Agamas are diurnal and relatively large with scaly, spiky bodies and large heads. Most male agamas have brightly colored heads. They can be found sunning on top of rocks in open areas. Skinks are fast-moving lizards living on the forest floor. Chameleons differ from other lizards in shape and size. They are slow-moving reptiles that live in trees and bushes. Chameleons are best known for the ability to blend into their environment. They can change color in different environments turning from green to brown or yellow to blend in with the surroundings. Along with the African seasonal forests, the dry forests of Madagascar house many species of chameleons.

The African seasonal forest is also home to amphibians, particularly frogs and toads. Tree frogs, rocket frogs, reed frogs, and bullfrogs are found in these forests. Many species of toads are also present. Toads vary in color; many can blend in well with the leaf litter or trees, making them almost invisible. True toads or bufonids have been successful in both Africa and South America. Africa is thought to be the place of origin of these toads, which later dispersed into the Americas. Several frogs and toads are endemic, with limited distribution within forested areas.

Insects and Other Invertebrates

Insects, arachnids, and crustaceans inhabit the seasonal forests of Africa. Beetles are a diverse order of insects in Africa with specialized niches. Many eat plants and others are associated with every kind of decomposing matter, while others are parasitic. New beetle species continue to be discovered as research is increased in these forests. Butterflies and moths are another group found in these forests. In the seasonal forests, butterflies and moths appear seasonally with the flowering of plants. Thousands of species are estimated to occur in the forests, although no current count is available. Some butterflies take in nectar while others choose fruit, dung, dead animals, and even animal perspiration as sources of nourishment. The butterfly families that are common in African forests are swallowtails, monarchs, brown butterflies, snout butterflies, the whites, the blues, and nymphs. None of these families is unique to Africa.

Termites are abundant in seasonal forests. These social insects play a crucial role in maintaining the ecosystems. Tropical Africa has the richest diversity of termites in the world, particularly soil-feeding termites. Their ability to feed on dead plant material makes them vital players in the forest. Termites are estimated to consume up to one-third of the annual litter, by decomposing it completely or making it more available for other decomposers. Termites are an important food source for some specialized forest mammals, such as pangolins and aardvarks. Some termites build large mounds or elaborate arboreal nests, others live entirely underground, often sharing the nest with pouched rats. Like ants, termites maintain large colonies of a few hundred to several million individuals. Three main groups of termites inhabit the seasonal forests: damp wood termites, dry wood termites, and the so-called higher termites (see Chapter 3).

Ants, bees, and wasps are found throughout the seasonal forests of Africa. Many have well-developed social structures and live in communities. Some feed on pollen and nectar, while others hunt small animals. Ants can be predators, decomposers, seed dispersers, and guardians of new growth. They can be found at all layers of the forest, using many different food resources. Army ants, driver ants, and weaver ants live within the forest. The African expression of the rainforest in Chapter 3 provides a detailed description of these types of ants. Other ants have important mutualistic relationships with trees, particularly acacias. While the ants protect the tree from herbivores, the tree provides the ants with shelter and food (see Figure 5.17). Insects such as grasshoppers, crickets, stick and leaf insects, cockroaches, mantids, flies, and fleas all play important roles, and each exploits a different niche within the forest.

Arachnids are numerous in the rainforest. Burrowing scorpions, crawling scorpions, flat scorpions, thick-tailed scorpions, and whip-tailed scorpions are found within the seasonal forests. Spiders are abundant nocturnal predators. Numerous other invertebrates play important roles in the maintenance of the tropical seasonal forests in Africa. Millipedes are scavengers and decomposers of vegetation and animals. Carnivorous centipedes feed on arthropods, worms, and small vertebrates.

Human Impact on the African Seasonal Forests

Humans have greatly affected the seasonal forests of Africa and caused the loss of most forests throughout the continent and on Madagascar. Increased populations, forest settlements, and clearing forests for agricultural and fuelwood are the main reasons for forest destruction. As population increases, further encroachment into the forest takes its toll on the plants and animals, making them vulnerable to degradation and loss. Most of the seasonal forests of West Africa have been destroyed already. What remains are small fragments interspersed amid the vast savannas. Decreased forested land leads to changes in climate, causing further desertification at forest margins. With large-scale human movements forced by armed conflict

Figure 5.17 Ants have a special relationship with acacias. They protect the tree from herbivory and in return the tree provides the ants with food and shelter. *(Photo by author.)*

and tribal wars, hundreds of thousands of people have migrated to the forest seeking protection or peace. Often out of necessity, these people are damaging the forest by hunting and clearing land. Trade in bushmeat is becoming a greater threat as exploited agricultural lands are no longer able to produce an economical source of protein. Meat from forest animals is the primary source of protein for people in rural areas as well as in some cities. Forestry operations often support the hunting of large game by providing access to forests and facilitating transport of meat from forest to cities.

On the Cape Verde Islands, the introduction of exotic species, such as rats, sheep, goats, cattle, and green monkeys have devastated the native flora and fauna. Overgrazing by livestock has resulted in extensive soil erosion, and the demand for wood has resulted in deforestation and desertification of much of the remaining dry forests on the islands. As noted earlier, the majority of the dry forests of Madagascar have been destroyed already. Very little forest area remains (estimates range from 4,500 to 8,000 mi^2 [12,000–20,000 km^2]), most in small fragments. The major threat to these dry deciduous forests is clearing and fragmentation. Expanding rural populations have increased the pressure on the remaining forested lands. Clearing the forest has endangered many of the lemur species. Several areas of dry forest are under protection in the Ankarafantsika National Park and the Tsingy de Bemaraha World Heritage site, while others are exploited.

Little of what is left of the seasonal forests of Africa is protected or set aside as reserves. Fortunately, much of the Zambezian dry evergreen and miombo forests remain largely undisturbed due to the absence of surface water and low human occupancy. An area of this forest is under protection in the West Liuwa National Park in Zambia near the border of Angola, although hunting is still a threat to wildlife in this area. Tanzania, Angola, Zambia, Mozambique, and the DRC have established reserves and national parks to protect small sections of seasonal forests and their inhabitants.

While much needed attention is placed on protecting African rainforests, little has been placed on protecting the seasonal forests. International environmental organizations abroad, as well as those in country, are now working to bring some protection to the last remnants of the seasonal forests in the African region.

Asian-Pacific Tropical Seasonal Forests

The Asian-Pacific expression of the Tropical Seasonal Forest Biome is also called the Indo-Malayan or Australasia Dry Forests and the Indo-Asian Monsoonal Forests. These forests are located in tropical latitudes, north and south of the tropical rainforest throughout India, Southeast Asia, Indonesia, and the islands of the Pacific. Seasonal forests are found in Thailand, Cambodia, Vietnam, Bangladesh, Laos, and Myanmar, and along the Decca Plateau in India and into Sri Lanka. Seasonal forests also occur on many of the islands of Indonesia including the Indonesian archipelago on both sides of Wallace's Line (see Figure 5.18). The islands of New Caledonia and Fiji have remnants of these tropical seasonal forests. The Asian-Pacific region is divided into two subregions: West and East Malesia. Wallace's Line (see Chapter 3) marks the boundary between the two subregions and the border of the Sunda and Sahul shelves.

Origins of the Asian-Pacific Rainforest

Africa, Europe, and Asia including West Malesia, share many taxa due to the long-term connections between the continents. The southeastern section of the region known as East Malesia is reflective of the separation of the larger Gondwanan landmass that left Australia isolated for millions of years. Australia and the

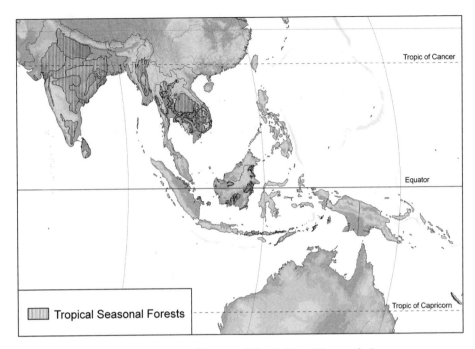

Figure 5.18 Map of Asian seasonal forests. *(Map by Bernd Kuennecke.)*

islands of the East Malesian subregion evolved their own distinctive flora and fauna found nowhere else on Earth. When the Australian plate moved northward and collided with the Asian continent (about 30–20 mya), dispersal of Asian species into the area was made possible, and the subregion became a mix of species from the two subregions. Some distinctive differences in the flora and fauna remain between the subregions. Wallace's Line between Borneo (Kalimantan) and Sulawesi (Celebes), and Bali and Lombok farther south, defines the end of the Sunda Shelf, a shallow extension of the continental plate under most of Indonesia. The deep trench at the end of the shelf has been a significant barrier to species dispersal, keeping species (particularly mammals) distinctive and separate. Fluctuations in sea level in the past exposed land bridges, allowing some exchange of species among islands. Today, plants and animals are transported throughout the region intentionally and unintentionally by humans. Aggressive invasive species have begun to dominate in certain areas, threatening many species and habitats.

Climate

The climate throughout the Asian-Pacific seasonal forests is significantly influenced by monsoonal cycles. Monsoons are seasonal shifts in wind directions in which moisture-laden air moves from ocean to land during the summer, and drier air moves from land to sea in the winter. Most of Southeast Asia, including the mainland, experiences a seasonal monsoon climate with a wet summer and dry winter.

During the Northern Hemisphere's winter, a high-pressure center over Tibet sends relatively cool, dry air to mainland Southeast Asia and northern Sumatra. During the summer, a center of high pressure over the Indian Ocean sends warm, moist air northeast over much of the same area. This creates a seasonal monsoon climate over much of mainland Southeast Asia, including Myanmar, Thailand, Cambodia, Laos, and Vietnam. The seasonal shifts in the ITCZ control the timing of the monsoons.

Annual rainfall in the Asian-Pacific seasonal forests is quite similar to that of the rainforest but concentrated while the sun is highest in the sky. Average temperatures range from 72°–87° F (22°–31° C), with the highest temperatures occurring just before the rainy season. Total annual rainfall reaches 40–70 in (1,000–1,800 mm), with higher totals on the windward sides of the mountains. During the rainy season, consistent cloud cover blocks incoming solar radiation, causing temperatures to decrease.

Changes in the monsoonal cycle can lead to prolonged drought. Periods of hot, dry weather can create an environment vulnerable to fire. Many of the seasonal forests are not resistant to such fires and can be converted to scrub and grasslands in an environment of high fire frequency. El Niño events can affect this region significantly, effectively weakening the monsoon and pushing it toward the Equator. This creates a period of prolonged drought, often leading to extensive forest fires.

Tropical cyclones (typhoons, hurricanes) affect seasonal forests between 10° and 20° N and S latitude. These include the forests of Bangladesh, the Philippines, and much of Melanesia and Australia. The occurrences of cyclones into forested areas create stands of disturbed forest, where fast-growing, sun-loving, often-invasive pioneer species tend to dominate.

Soils

The soils of the Asian-Pacific seasonal forests are similar to those of the rainforest, but the percentage of area covered by each soil type differs significantly. Deeply weathered, infertile oxisols make up a large percentage of soils in the seasonal forest of the Asian and Indian mainland. The majority of soils on older islands in the region are ultisols. Other forests soil types on more recent volcanic islands include inceptisols, alfisols, and entisols. Most forests with these soils have been used for agricultural production, as they tend to be richer, well-drained soils of volcanic or alluvial origin.

Vegetation in the Asian-Pacific Tropical Seasonal Forests

Like the rainforest, floristically this region is called Malesia and is separated into two more or less distinctive subregions. West Malesia includes India, Southeast Asia, the Philippines, the Malayan Archipelago, Brunei, and the islands of

Indonesia. East Malesian includes the area east of Wallace's Line: Sulawesi, Lombok, New Guinea, the tropical islands of Melanesia, and northeastern Australia.

Seasonal forests, often called monsoonal forests in this region, border tropical rainforests. This boundary is largely determined by length and severity of the dry season as it relates to soil moisture availability. Several types of seasonal forests exist within the Asian-Pacific region. They include moist deciduous forests, dry deciduous forests, and mixed deciduous forests. The region also contains dry evergreen forest and dry thorn forest, which are all considered part of the Tropical Seasonal Forest Biome. Different forest types vary in terms of forest structure and dominant plant species.

Forest Types and Structure

Tropical moist deciduous forests are the wettest of the seasonal forests found in this region. Rainfall averages 40–50 in (1,000–1,800 mm) or more annually, with a distinct dry season. Eighty percent of the total precipitation falls during the summer, and a distinct dry season of four to six months occurs during the winter months.

The moist deciduous forest can have a closed or open canopy. Taller trees reach heights of 33–130 ft (10–40 m), significantly smaller than those of the tropical rainforests (see Figure 5.19). In the lowland forests, climbing vines and lianas are less abundant. The understory contains many different palms as well as smaller trees. As elevation rises to 655–1,300 ft (200–400 m), vines and lianas become more common. In India, Thailand, and Myanmar teak (Verbenaceae) and sal trees (Dipterocarpaceae) dominate on the less fertile soils, while trees in the legume family (Caesalipiniacea subfamily) dominate on more fertile soils.

Mixed deciduous forests are found primarily in the northwest of Thailand and into Myanmar and India. The structure of the mixed deciduous forest is complex

Figure 5.19 Forest structure in monsoon forest. *(Illustration by Jeff Dixon.)*

and, as its name implies, the species present are diverse. This type of tropical seasonal forest has a high, closed canopy layer at 80–100 ft (25–30 m) or higher, and a relatively open understory of small trees, and bamboo reaching above 23 ft (7 m) in height with few epiphytes and lianas. The forest floor tends to be open with little low-growing vegetation. The dominant tree in this community is also teak. Other important trees include *Xylia kerrii* (locally known as daeng), tropical almond, and a species of crape myrtle (loosestrife family). Diversity of tree species is high. Mixed deciduous forests dominated by teak have been heavily disturbed due to logging this valuable wood.

Dipterocarp forests are the most extensive deciduous forest type, covering more area in northeast India, Myanmar, and Thailand, as well as Laos, Cambodia, and Vietnam, than any other forest type. This dipterocarp forest is characterized by 40–60 in (1,000–1,500 mm) of highly seasonal annual rainfall, and a five- to seven-month dry season. Most areas where dipterocarp forests occur are mountainous or hilly and are dry with shallow soils. Most trees are leafless between February and April. Fires commonly occur in this forest type between December and March at intervals of one to three years. The structure of the deciduous dipterocarp forest can range from a closed canopy to more open woodland. The canopy is typically low at 16–26 ft (5–8 m) in height with emergents reaching 33–40 ft (10–12 m). The deciduous dry forests in Thailand, Laos, Cambodia, and Vietnam contain all six deciduous dipterocarp species. Four species including taengwood, dark red meranti, *Dipterocarpus obtusifolius,* and *D. tuberculatus* (no common names) dominate. Smaller trees in this forest are from the legume family and include *Sindora siamensis* (locally called tete hoho) and *Xylia xylocarpa* (pyinkado). The leaves of these trees tend to be large and thick. Cycads and palms are common in the understory. Where the forest is more open, grasses dominate the understory.

Other dry deciduous forests in India are located on the geologically ancient Decca Plateau. They include woodlands with an open canopy dominated by two members of the legume family (Fabaceae), *Hardwickia binata* and *Albizia amara,* and having an upper canopy of 50–80 ft (15–25 m) and an understory 32–50 ft (10–15 m) tall.

Dry evergreen forests contain a mix of deciduous and evergreen trees. The evergreen trees tend to have smaller leaves that are leathery. The forest canopy can be quite tall at 82–98 ft (25–30 m), with a mix of dipterocarps from the wet forest. The lower canopy can range from 23 to 75 ft (7–23 m) in height. The understory of these forests is rich in lianas. Sri Lanka has a large remnant of evergreen forests that are important habitat for the Asian elephant.

The seasonal forests of the Asian-Pacific region are relatively poor in climbing species apart from palms and dipterocarps. However, some climbing species in the dogbane, milkweed, morning glory, legume, and squash families are present. Spiny rattan palms are the most diverse group.

Seasonal forests of the East Malesian subregion contain a mix of species from Australia and Asia, but only fragments of seasonal forests in this subregion still exist. These forests include moist deciduous forests, dry deciduous forest, dry semi-evergreen and evergreen forests, as well as dry thorn forests. Seasonal forests are found on the volcanic islands of Lombok, Sumbawa, Flores, Komodo, Alor, and Timor along the Indonesian archipelago. They also occur on the islands of New Caledonia in the Pacific.

Dipterocarps are present, but they play a less dominant role. In the moist deciduous forest, tamarind and wild almond are the dominant trees. In Timor, Burmese rosewood is a dominant tree. In the dry deciduous forests, Ceylon oak (Sapindaceae), East Indian rosewood (Gesneriaceae), and trees in the frankincense family are abundant, with many lianas in the orchid genus (*Bauhinia*) and euphorb family. Semi-evergreen species include native eucalyptus and sandalwood, both valuable timber trees. Many species are endemic to this area. Primitive conifers of the araucaria (Araucariacea) and podocarp (Podocarpaceae) families have genera that are limited in the tropics to the islands of the Indonesian archipelago east of Wallace's Line and in Australia and New Guinea.

The dry forests on the islands of New Caledonia contain a large number of endemic plants. These islands were once part of the larger Gondwana landmass and broke off from Australia 85 mya. Their isolation in the Pacific created a unique environment. Along the west side of the island, dense forests of Gaic (acacia) trees dominate with trees from the coffee family, including the genera *Canthium, Gardenia*, and *Pittosporum*. The forest is thick with vines, understory shrubs, and grasses. The dry forests in New Caledonia are highly vulnerable to extinction because many plants on the islands occur there and nowhere else. A destructive widespread fire could wipe out many species.

The islands of Fiji once had abundant deciduous forests, but they have been reduced from 2,900 mi^2 (7,557 km^2) to less than 38 mi^2 (100 km^2). These remnant forests contain fragrant sandalwood and several conifers in the podocarp family along with cycads and bamboo.

Animals of the Asian-Pacific Tropical Seasonal Forests

A diverse assemblage of wildlife is associated with the seasonal forest of the Asian-Pacific region. Mammals found in these forests are similar to those found in the rainforest nearby with few exceptions. An abundance of bird species are present, but like the other dry forest regions, few are endemic. Reptiles and amphibians are similar to those of the rainforest, but the number of species is lower, particularly for those that require permanent water sources. Hundreds of invertebrate species can be found within these forests, from very small ants and beetles to the very large bird-eating tarantulas and Asian forest scorpions.

Species composition is often dependent on the duration of dry season, along with the particular type of forest present. The extent of degradation and fragmentation also influences the composition of species and the number of individuals in the forest. Larger intact forests hold a greater diversity as well as larger populations than smaller fragmented forests.

Mammals

Many of the mammals found in the surrounding rainforests take advantage of the abundance of flowers and fruit within the tropical seasonal forest at the start of the dry season. Some migrate to the deciduous forest and return to the rainforest as the dry season intensifies. Many of the mammals found within the West and East Malesian regions have been discussed in Chapter 3. The complex structure, variety of habitats, and diversity of leaves, flowers, fruits and seeds support large numbers of mammal families (see Table 5.4).

Specialized insect eaters include pangolins, shrews, and tree shrews. Terrestrial pangolins are present in the seasonal forests of the West Malesian subregion. Several types of Asian shrews live in the forests and woodlands of the region. Shrews tend to be small and feed frequently on invertebrates. Several endemic species are found on the islands of Indonesia. Tree shrews are not true shrews and spend most of their time foraging on the forest floor. Tree shrews resemble ground squirrels with bushy tails and pointed snouts. They are found in West Malesia but are absent in East Malesia.

Bat diversity is high in the seasonal forests of the Asian-Pacific region. Bats in both suborders, Megachiroptera and Microchiroptera, are abundant. Microchiroptera make up the majority of bats in the Asian-Pacific region. They are mainly insect-eating bats that use echolocation to find their prey. Most of the bats in the region are aerial insectivores, catching insects as they fly. The East Malesian subregion has bats similar to those found in the rest of the region.

Rodents of the Asian-Pacific seasonal forests include squirrels, Old World porcupines, Old World mice and rats, and bamboo rats. Squirrels are particularly abundant in the West Malesia subregion of these forests. Most are arboreal, with a diet of fruits, seeds, leaves, and insects.

Old World porcupines are large, slow-moving mammals that rely on their imposing quills rather than on speed or agility for defense. They are found in West Malesia. These mammals are terrestrial and excellent diggers, live in burrows, and consume many kinds of plant material, as well as carrion.

Rats and mice are less abundant in the Asian-Pacific rainforest. Most are terrestrial. Mice tend to feed on fruits, seeds, and grasses, while rats also eat insects, mollusks, or crabs. Old World mice and rats are present in both subregions, with several endemic species found on the islands of East Malesia. Sri Lanka is home to the endemic Ceylon spiny mouse and the Nolthenus long-tailed climbing mouse. Rajah rats, the favorite prey of leopard cats in Thailand, are found in abundance in the deciduous dipterocarp forests. Bamboo rats are a small family of rodents that live in the West Malesia subregion. They are fossorial, but come to the surface to forage. Bamboo rats are most commonly found in bamboo forests and agricultural lands.

Table 5.4 Asian-Pacific Mammals Found in the Seasonal Forests

Order	Family	Common Names
Marsupials (Metatheria)		
Dasyuromorphia	Dasyuridae	Antechinus, quolls
Diprorodontia	Phalangeridae	Brushtail possums
	Macropodididae	Wallabies
Peramelemorphia	Peramelidae	Bandicoots
Placental Mammals (Eutheria)		
Artiodactyla	Suidae	Pigs
	Cervidae	Muntjacs, sambar, axis deer
	Boviidae	Buffalo, banteng, gaur
Carnivores	Caniidae	Jackals, wild dogs
	Felidae	Leopards, tigers
	Herpestidae	Mongoose
	Mustelidae	Otters and ratels
	Viveridae	Civets and genets
	Ursidae	Sloth bear, sun bear
Chiroptera	Craseonycteridae	Bumblebee bat and hog-nosed bats
	Emballonuridae	Sac-winged bats, sheath-tailed bats, and relatives
	Hipposideridae	Leaf-nosed bats
	Megadermatidae	False vampire bats
	Molossidae	Free-tailed bats
	Nycteridae	Slit-faced bats
	Pteropodidae	Fruit bats
	Rhinolophidae	Horseshoe bats
	Rhinopomatidae	Mouse-tailed bats
	Vespertilionidae	Insect-eating bats
Perissodactyla	Tapiridae	Tapirs
	Rhinocerotidae	Rhinoceros
Pholidota	Manidae	Pangolins
Primates	Cercopithecidae	Macaques, langurs, leaf monkeys, proboscis
	Galagidae	Galagos
	Hylobatidae	Gibbons, lesser apes
	Lorisidae	Lorises
Proboscidea	Elephantidae	Elephants
Rodents	Hystricidae	Old World porcupine
	Muridae	Old World rats and mice
	Rhizomyidae	Bamboo rats
	Sciuridae	Squirrels
Scandentia	Tupaiidae	Tree shrews
Soricomorpha	Scoricidae	Shrews

Primates are abundant in the West Malesian subregion and absent in the East Malesian subregion. Primates live permanently or seasonally in these seasonal forests; they include lorises, gibbons, langurs, macaques, and leaf monkeys. Orangutans, once abundant, are no longer present due to habitat fragmentation, hunting, and deforestation.

Asian elephants are the largest mammals in the forests of West Malesia. Asian elephants are smaller than African elephants with smaller ears and tusks. They tend to be nocturnal and feed on fruits from the forest like wild bananas, bamboo, and other vegetation. They spend their days resting deep inside the forest. Asian elephants from nearby rainforests or savannas will migrate into the seasonal forests when fruit is available. The dry forests of Sri Lanka hold the largest Asian elephant population in the region.

Of the two rhinoceros species in the Asian-Pacific region, only the two-horned Sumatran rhinoceros is thought to still be present in the seasonal forest. Populations continue to decline as habitat disappears.

Other ungulates of the seasonal forests include the Indian axis deer, spotted deer, blackbuck antelopes, and chousinga (a small four-horned antelope) of India, as well as sambars, barking deer, gaurs (wild ox), wild water buffalo, serows, Eld's deer, and the endangered kouprey and banteng in the forests of Southeast Asia. Wild pigs and bearded pigs are found in both forested areas and in nearby agricultural areas.

The carnivores that live in the deciduous forests include the smaller carnivores—such as mongooses, civets, palm civets, linsangs, leopard cat, golden cats, Asian wild dogs (dholes), and the Asian jackal—and the larger cats, clouded leopards, common leopards, and tigers, as well as sun bears, sloth bears. The Himalayan black bear lives in the dry forests of Myanmar. None of these carnivores are found east of Wallace's Line. Only a few marsupial carnivores live in the seasonal forests of East Malesia, including marsupial mice and quolls.

Asian wild dogs (dholes) are found in the seasonal and scrub forests of India, Myanmar, and the Malaysian Archipelago. They tend to hunt in packs killing wild pigs, deer, and an occasional monkey. Of the 10 subspecies of dholes in existence, four are considered threatened or endangered. The Asian or golden jackal is more widespread, occurring in southern Asia as well as Africa and southeastern Europe. Jackals tend to live and hunt in small family groups with large populations scattered throughout their distribution.

The largest of the Asian cats in the forest are spotted leopards and tigers. The spotted leopard is a large nocturnal cat that rests in the forest trees during the day. Their diet varies from monkeys and ungulates to rodents, birds, and rabbits. Tigers are still present in the West Malesian subregion of the Asian Pacific, although their numbers are dwindling (see Plate XVI).

Sun bears and sloth bears are found in the West Malesian subregion. Sun bears live in the tropical forests of southern Asia, Myanmar, Malaysia, and Sumatra. They spend most of their time in trees eating lizards, birds, fruit, ants, termites, and honey. The sloth bear lives in the forest of southern India and Sri

....

The King of the Dry Forests

Tigers are extremely powerful large cats. They
are nocturnal solitary hunters that defend large
territories while searching for food. Tigers communicate vocally with roars, growls, snarls,
grunts, moans, mews, and hisses. Each sound
can reflect the tiger's intent or mood. Three
subspecies of tiger are found in the seasonal
forests of the region: Bengal, Indonesian, and
Sumatran. Bengal tigers are found in India and
Bangladesh. The Nigarjunasgar Tiger Reserve in
central India is one of the largest and most important protected reserves for the Bengal tiger.
Indochinese tigers are found in Cambodia,
Laos, Malaysia, Myanmar, Thailand, and Vietnam. Sumatran tigers, the smallest of all tigers,
are found only on the Indonesian island of
Sumatra. Bali and Javan tigers are thought to be
extinct. Tiger populations are severely impacted
by deforestation and loss of habitat as well as
illegal poaching and trade. Many tigers are killed
for their fur as well as for tiger products used for
traditional medicinal preparations.

....

Lanka, and feeds largely on termites and bees.
Bear populations are in steep decline due to loss
of habitat and hunting. The Himalayan or Asiatic black bear are seasonal visitors to the dry forests of Myanmar, Thailand, and Vietnam when
fruit is plentiful. During the winter, they descend
from the mountains to the tropical forests where
food is plentiful.

Eastern Malesia mammal fauna are a mix of
Australian and Asian species. As most of the dry
forests in this subregion occur on islands, fewer
mammals are present. Several endemic species
inhabit the dry forests of these islands, they
include several endemic rats; endemic bats,
including flying foxes, long-eared bats, and horseshoe bats; and several endemic shrews on
Komodo and Timor. Marsupials in the seasonal
forests of East Malesia fill similar niches as placental mammals in the western subregion of the
Asian Pacific. They live in the trees and on the
ground, and are herbivores, omnivores, and carnivores. The isolation of this subregion from the rest
of the region allowed ancient marsupials to
evolve. Their isolation and lack of large predators
are probably responsible for their presence today.

Birds

Birds are abundant in the seasonal forests of the Asian-Pacific region. Many families in this region are shared with Africa, most likely due to dispersal during intermittent forested connections. Some of the important families shared between the
two regions include hornbills, bulbuls, and sunbirds. Hornbills are large birds with
black, white, and yellow feathers and huge bills. They are common in the open seasonal forests. Hornbills eat almost anything, although a few are strictly carnivorous, while others are frugivores.

Many ground-dwelling birds including pheasants, partridges, tragopans, firebacks, great and crested argus, and peacock pheasants forage the forest floor and
shrub layer for food. Doves and pigeons are abundant on the forest floor and the
lower canopy levels. In the trees, myna birds are common, along with brightly colored sunbirds, flowerpeckers, spiderhunters, and honeyeaters. Like the sunbirds of
Africa, Asian sunbirds are small birds with long curved bills. They are mainly nectar feeders and can be bright yellow, red, purple, and olive green. Many of these

birds are seasonal visitors to the forest, returning to the rainforest when flowers and fruit are no longer available.

Bulbuls, Old World Warblers, thrushes, flycatchers, shrikes, and babblers are insectivorous birds of the seasonal forests found traveling in flocks of mixed species. They all tend to be small and drab brown in color. Different species search for insects on different parts of a plant, such as the leaves, twigs, or tree trunks. Other forest birds include starlings, robins, drongos, woodpeckers, piculets, and barbets. The predatory birds of the forest include osprey, fish eagles found along waterways, and eagles, buzzards, hawks, kites, and falconets. They are all excellent hunters with diversified diets of fish, reptiles, small rodents, and other small mammals.

In the Australasia subregion, birds have diverged greatly in appearance and behavior. A number of endemic ground doves, pigeons, parrots, lorikeets, honeyeaters, figbirds, and thrushes are present on the eastern Indonesian archipelago. Parrots and lorikeets are more abundant in the East Malesian subregion than in the west.

Reptiles and Amphibians

Snakes, lizards, crocodiles, turtles, and a vast array of frogs, toads, and caecilians inhabit the seasonal forests. They are terrestrial, arboreal, fossorial, or aquatic. More than a hundred different tropical snakes are found in the Asian-Pacific region with less than 10 percent considered poisonous. Poisonous snakes include cobras, vipers, and adders. Two cobra species, the king and the Indian cobra, are found in this region. The king cobra is the most deadly and is much larger than other cobras. Its venom is a potent neurotoxin that affects the nerve and respiratory systems. The Indian cobra is a medium-size snake that feeds on rodents, lizards, and frogs. Its venom damages the nervous system of the prey, paralyzing and often killing it. In East Malesia, several other venomous snakes can be found. They include the red-bellied black snake, the highly venomous eastern brown snake, and the less venomous brown tree snake.

Some of the venomous vipers in the forests include the Malayan pit viper; hundred pace viper; hump-nose and palm vipers; and the South Asia, bamboo, and temple pit vipers.

Nonpoisonous snakes in the Asian-Pacific region include pythons and tree snakes. The large reticulated python, Indian python, and Timor python along with green tree pythons are present in the region. Australia is also home to the carpet python and amethystine python. Tree snakes are smaller and often beautiful snakes that live in the trees eating birds, eggs, small arboreal mammals, and reptiles. They are fast and expert climbers. Their coloration, similar to leaves and bark, provides good camouflage.

Ground-dwelling snakes include racers, rat snakes, keel backs, pipe snakes, and burrowing blind or worm snakes. Kukri snakes, reed snakes, little brown snakes, slug snakes, wolf snakes, and mock vipers are other common forest dwellers in the Asian-Pacific region.

A great diversity of lizards live in the region, with more than 150 species recorded. Agamid lizards, monitors, geckos, and skinks are present in both subregions of the rainforest. Agamid lizards are common throughout the region. They are the Old World equivalent to iguanas and live in trees, on the ground, and along waterways. The largest lizards of the forest are the varanids or monitor lizards. The varanids are an ancient group of lizards found in Africa and throughout southern Asia, the Indonesian islands, and Australasia. These lizards are strong, diurnal lizards ranging in length and weight from the short-tailed monitor at about 8 in (200 mm) and 0.7 oz (20 g), to the Komodo dragon at 10 ft (3 m) and 120 lb (54 kg). The Komodo dragon is the world's largest lizard. It is restricted to the islands of Komodo, Flores, and the smaller islands in the eastern part of the Indonesian archipelago. Smaller monitors are found throughout the entire region. Skinks are the most diverse group of lizards, with the greatest numbers occurring in the Asian-Pacific region. They are slender, fast carnivorous lizards that eat invertebrates and small rodents. Tree skinks, sun skinks, brown skinks, and slender skinks live in the dry forests of the Asian Pacific. Geckos are small insectivorous lizards found in warm climates throughout the world. Golden geckos, house geckos, Indo-Pacific geckos, tree geckos, and Tokay geckos all inhabit this region.

Toads, frogs, and caecilians are the amphibians of the Asian-Pacific seasonal forest. Many live in or near water, while others spend their lives in the trees or forest floor. True toads—including the Sulawesi toad, Asiatic toad, forest toad, and four ridge toad—and true frogs—such as field and creek frogs, puddle frogs, cricket frogs, rock frogs, the Malaysia frog, and rhinoceros frog—are found throughout the region. Litter frogs and horned frogs live on the forest floor among the decomposing leaves, and bullfrogs, chorus frogs, black-spotted frogs, stick frogs, and narrow-mouthed frogs emerge from burrows after it rains. The yellow-striped caecilian is a legless amphibian, worm-like in appearance that lives underground in the seasonal forest.

Insects and Other Invertebrates

Like other regions, the seasonal forests in the Asian-Pacific region host a multitude of insects and other invertebrates that play important roles in the forest. Insects are the largest class of invertebrates in the forest. Butterflies, moths, ants, wasps, bees, termites, beetles, and stick and leaf insects are incredibly varied and have developed unique adaptations to survive and prosper.

Butterflies are abundant during certain times of the year. Many of the butterflies in this region are in the five main butterfly families: birdwings and swallowtails, milkweed butterflies, gossamer-winged butterflies, satyrs, wood and tree nymphs, and saturns. The birdwings and swallowtails are some of the largest and most spectacular butterflies in the world.

Moths are more numerous than butterflies, but are less well studied. Most moths are active at night; however, swallowtail moths are active during the day. Hawk moths, sphinx moths, and hornworms are common in the seasonal forests.

These moths are noted for their flying ability, especially their ability to move rapidly from side to side while hovering.

Mantids, stick insects, and leaf insects are abundant in the forests of the Asian-Pacific region. They closely resemble twigs and leaves. These insects are primarily nocturnal and avoid detection during the day by remaining still. Mantids use camouflage to avoid detection by the prey they seek to capture. They lie in wait to ambush an unsuspecting insect. Stick and leaf insects and mantids are quite successful in employing camouflage strategy.

Termites are abundant throughout the Asian-Pacific forests and are the dominant decomposers of the seasonal forests. Damp wood termites feed mainly in fallen trees. This family of termites is thought to have originated in this region. The majority of termites in the forest are in the higher termite family. As noted in the African region, these termites are divided into four subfamilies based on feeding and defense strategies. Soil-feeding and wood-feeding termites belong to this family. Most live in nests on the forest floor or underground, although some make nests in the trees. In addition to playing a major role in the decomposition of fallen wood, and recycling nutrients within the forest, termites are a major source of food for pangolins, shrews, sun bears, and sloth bears.

Ants are social insects abundant in the seasonal forest. They have been discussed at length in the previous chapters. Bees, wasps, and hornets live in the forest but are less abundant here than in the rainforests. Bees in seasonal forests must be able to adapt to long intervals between flowering events, which is why less are found here than in the rainforests. Social bees play a major role in the pollination of many dipterocarp species in Southeast Asia.

Spiders, scorpions, centipedes, millipedes, and other invertebrates are common in seasonal forests. Spiders can be classic web weavers, trap-door spiders, and those that sit and wait to ambush an unsuspecting victim. Very large tarantulas, such as bird-eating spiders, wait in holes or crevices ready to ambush large insects. Scorpions are active hunters that attack large insects at night and spend their days under stones, bark, or in rotting wood. Whip scorpions are another invertebrate predator of the forest that feed on worms, slugs, and other arthropods. Centipedes and millipedes are both common forest creatures. Centipedes are nocturnal predators that feed on other invertebrates. Millipedes are generally active during the day and feed on soft decomposing plant matter.

The animals of the Asian-Pacific seasonal forests are diverse and abundant. They have developed adaptation strategies to survive in areas where food is only seasonally abundant. Birds are plentiful throughout the region and, like mammals, have become feeding specialists or developed behavioral strategies that allow them to inhabit particular niches within the forest. Seasonal migration is common among many mammals and birds that inhabit these forests. Invertebrates make up the majority of forest animals, some having developed unique appearances to mimic vegetation, others traveling in large social groups working together to find food and defend their homes.

The animals and plants of the Asian-Pacific Seasonal Forest Biome live in close relationships with each other and the physical environment. Changes in rainfall or temperature can greatly affect their survival. Conversion, degradation, and fragmentation exacerbate the decline of tropical seasonal forests. Continued increases in human population, large-scale agriculture expansion, unsustainable forestry, and illegal poaching of trees and animals are the main reasons for the loss of the tropical seasonal forests in the Asian-Pacific region.

Human Impact on the Asian-Pacific Seasonal Forests

The Tropical Seasonal Forest Biome is the most threatened of all biomes, and the forests of the Asian-Pacific region are becoming increasingly vulnerable. In some parts of this region, up to 90 percent of the forest has been destroyed. In Thailand and Vietnam, more than 65 percent of the forest has been cleared for shifting cultivation, large- and small-scale legal and illegal logging, and conversion for long-term agricultural goals. Seventy-five percent of the evergreen dry forests in Sri Lanka have been deforested for agriculture, resettlement, and small-scale logging.

In Southeast Asia, clearing the forest for cash crops such as cotton, rice, teak farms, and fruit orchards threatens the remaining forests. Other areas have been cleared for rubber tree, coffee, and tea plantations. Other threats include the exploitation of valuable hardwood trees (like teak) and other plant resources, and rampant hunting to supply the huge market demand in Vietnam and China.

In India, the primary threats to the remaining seasonal forests are from quarries, coal mines, large-scale clearing for agriculture, and hydroelectric projects. Shifting cultivation and the local communities' dependence on forest products continue to degrade the ecological integrity of the forests. Illegal timber operations are unlawfully removing trees in remaining forests at a rapid rate. In other areas where the human population is increasing rapidly, urbanization, industrialization, and agriculture associated with this growing population pose serious threats to the remaining forest fragments. Small, protected areas become vulnerable to these disturbances, and restoration of degraded areas around these small preserves is needed to maintain the integrity of these forest remnants.

Seasonal forests are dry for as much as half of the year and can be highly susceptible to fire. Frequent fires from intentional and unintentional burning can lead to total loss of the forest, causing desertification and conversion into fire-tolerant, less species-rich scrublands or dry grasslands.

The tropical seasonal forests of Asian-Pacific region contribute to worldwide biodiversity. They are vital for the survival of many animals. Seasonal forests remain largely insufficiently researched. Their rapid destruction presents many problems in terms of their influence of climate change locally and species extinction globally.

Further Readings

Books

Bullock, S. H., H. A. Mooney, and E. Medina, eds. 1995. *Seasonally Dry Tropical Forests.* Cambridge: Cambridge University Press.

Burgess, N., Jennifer D'Amico, Emma Underwood, Eric Dinerstein, David Olson, Illanga Itoua, Jan Schipper, Taylor Ricketts, and Kate Newman. 2004. *Terrestrial Ecoregions of Africa and Magascar: A Conservation Assessment.* Washington, DC: Island Press.

Hopkins, B. 1966. *Forest and Savanna.* London: Heinemann Press.

Janzen, H. D. 1988. "Tropical Dry Forest: The Most Endangered Major Tropical Ecosystem." In *Biodiversity,* ed. E. O. Wilson and F. M. Peter, 130–137. Washington, DC: National Academy Press.

Pennington, R. Toby, Gwilym P. Lewis, and James A. Ratter, eds. 2006. *Neotropical Savannas and Seasonally Dry Forests: Plant Diversity, Biogeography and Conservation.* Boca Raton, FL: CRC Taylor and Francis.

Robichaux, R. H., and D. A. Yetman, eds. 2000. *The Tropical Deciduous Forest of Alamos: Biodiversity of a Threatened Ecosystem in Mexico.* Tucson: University of Arizona Press.

Yamada, I. 1997. *Tropical Rain Forests of Southeast Asia: A Forest Ecologist's View.* Honolulu: University of Hawaii Press.

Internet Source

National Geographic and World Wildlife Fund. 2001. "Wild World, Terrestrial Ecoregions of the World." http://www.nationalgeographic.com/wildworld/terrestrial.html.

Appendix

Common and Scientific Names of Species in Tropical Seasonal Forests

Plants

Baobab	*Adansonia digitata*
Barriguda	*Larimus breviceps*
Blackwood	*Dalbergia melanoxylon*
Candelabra cactus	*Euphorbia trigona*
Ceylon oak	*Schleichera oleosa*
Daeng	*Xylia kerrii*
Dark red meranti	*g. Shorea*
East Indian rosewood	*Dalbergia latifolia*
Flamboyant tree	*Delonix regia*
Kapok tree	*Ceiba pentandra*
Madagascar palm	*Pachypodium lameri*
Miombo	*Brachystegia* spp.
Mnondo	*Julbernardia globiflora*
Palo barrocho	*Chorisia* spp.
Taengwood	*Shorea obtusa*
Teak	*Tectona grandis*

Mammals

Aardvark	*Orycteropus afer*
African elephant	*Loxodonta africana*
African golden cat	*Profelis aurata*
African wild dog	*Lycaon pictus*
Asian elephant	*Elephas maximus*
Asian jackal	*Canis aureus*
Asian wild dogs	*Cuon alpinus*
Banteng	*Bos javanicus*
Barking deer	*Muntiacus muntjak*

Bearded pig	*Sus barbatus*
Black howler	*Alouatta caraya*
Black rhinoceros	*Diceros bicornis*
Black-shouldered opossum	*Caluromysiops irrupta*
Black-tailed silvery marmoset	*Callithrix argentata*
Blackbuck antelope	*Antilope cervicapra*
Blue duiker	*Cephalophus monticola*
Brown brocket deer	*Mazama gouazoubira*
Brown lemur	*Eulemur fulvus*
Buffalo	*Syncerus caffer*
Bushy-tailed opossum	*Glironia venusta*
Cacomistles	*Bassariscus sumichrasti*
Caracal	*Caracal caracal*
Chacoan peccary	*Catagonus wagneri*
Cheetah	*Acinonyx jubatus*
Chousinga	*Tetraceros quadricornis*
Clouded leopard	*Neofelis nebulosa*
Collared peccary	*Pecari tajacu*
Eld's deer	*Rucervus eldii*
Fossa	*Fossa fossana*
Four-eyed opossum	*Metachirus nudicaudatus*
Gambian pouched rat	*Cricetomys gambianus*
Gaur	*Bos frontalis*
Giant anteater	*Myrmecophaga tridactyla*
Giant tuco tuco	*Ctenomys conoveri*
Golden-crowned sifaka	*Propithecus tattersalli*
Greater kudu	*Tragelaphus strepsiceros*
Hairy armadillo	*Chaetophractus villosus*
Himalayan black bear	*Ursus thibetanus*
Impala	*Aepyceros melampus*
Indian axis deer	*Axis axis*
Jaguar	*Panthera onca*
Jumping rat	*Hypogeomys antimena*
Kouprey	*Bos sauveli*
Leopard	*Panthera pardus*
Leopard cat	*Prionailurus bengalensis*
Lesser mara	*Pediolagus salinicola*
Lion	*Panthera leo*
Long-nosed armadillo	*Dasypus hybridus*
Long-tailed weasel	*Mustela frenata*
Maned wolf	*Chrysocyon brachyurus*
Mantled howler monkey	*Alouatta palliata*
Milne-Edwards sportive lemur	*Lepilemur edwardsi*
Miombo genet	*Genetta angolensis*

(*Continued*)

Mongoose lemur	*Eulemur mongoz*
Nine-banded armadillo	*Dasypus novemcinctus*
Northern naked-tailed armadillo	*Cabassous centralis*
Orangutan	*Pongo pygmaeus*
Perrier's sifaka	*Propithecus perrieri*
Puma	*Puma concolor*
Rajah rat	*Maxomys surifer*
Red howler monkey	*Alouatta seniculus*
Red river hog	*Potamochoerus porcus*
Rock hyrax	*Procavia capensis*
Sambar	*Rusa unicolor*
Serows	*Capricornis sumatraensis*
Serval	*Leptailurus serval*
Six-banded armadillo	*Euphractus sexcinctus*
Sloth bear	*Melursus ursinus*
South American gray fox	*Lyca lopex griseus*
Southern naked-tailed armadillo	*Cabassous unicinctus*
Spiny pocket mouse	*Liomys irroratus*
Spotted deer	*Rusa timorensis*
Spotted hyena	*Crocuta crocuta*
Striped hyena	*Hyaena hyaena*
Sumatran rhinoceros	*Dicerorhinus sumatrensis*
Sun bear	*Helarctos malayanus*
Tiger	*Panthera tigris*
Water buffalo	*Bubalus bubalis*
Western forest rat	*Nesomys audeberti*
Western forest mouse	*Macrotarsomys bastardi*
White-faced capuchin	*Cebus capucinus*
White-lipped peccary	*Tayassu pecari*
White-tailed deer	*Odocoileus virginianus*
Yellow-backed duiker	*Cephalophus silvicultor*

Birds

Black-headed Vulture	*Coragyps atratus*
Böhm's Flycatcher	*Muscicapa boehmi*
Chaco Chacalaca	*Ortalis canicollis*
Greater Rhea	*Rhea americana*
Guira Cuckoo	*Guira guira*
Little Thornbird	*Phacellodomus sibilatri*
Miombo Gray Tit	*Melaniparus griseiventris*
Miombo Rock Thrush	*Monticola angolensis*
Secretary Bird	*Sagittarius serpentarius*

Reptiles and Amphibians

American crocodile	*Crocodylus acutus*
Amethystine python	*Morelia amethystina*
Angola ornate frog	*Hildbrandtia ornatissima*
Angonoka tortoise	*Geochelone yniphora*
Bocage's horned adder	*Bitis heraldica*
Chacoan burrowing frog	*Chacophrys pierotti*
Fer-de-lance	*Bothrops asper*
Huila tree frog	*Leptopeltis anchietae*
Hump-nose viper	*Hypnale hypnale*
Hundred pace viper	*Deinagkistrodon acutus*
Indian cobra	*Naja naja*
King cobra	*Ophiophagus hannah*
Komodo dragon	*Varanus komodoensis*
Malayan pit viper	*Calloselasma rhodostoma*
Puff adders	*Bitis arietans*
Quill-snouted snake	*Xenocalamus bicolor*
Rock python	*Python sebae*
Timor python	*Python timoriensis*
Yellow-striped caecilian	*Ichthyophis kohtaoensis*

Invertebrates

Burrowing scorpion	*Opistophthalmus* spp.
Flat scorpions	*Hadogenes paucidens*
Whip scorpion	*Mastigoproctus giganteus*

6

Conclusion

Tropical forests have the greatest biodiversity of all biomes in the world, but now occupy less than 7 percent of the Earth's surface. Their position on and near the Equator with abundant sun and moisture produces a rich array of flora and fauna. Tropical forests are found in Central and South America, West and Central Africa and Madagascar, and most of Southeast Asia, India, Australia, New Guinea, and the islands of the Pacific. Each region has a unique set of plants and animals that have evolved over millions of years. Each region's flora and fauna bear remarkable similarities in size, shape, or behavior to those in the other regions; however, in most cases, they are unrelated. Similar environments have resulted in convergent evolution.

Tropical forests are the most ancient, diverse, and ecologically complex of the planet's ecosystems and support more than half of the Earth's species. They are a complex network of interrelated systems. Tropical forests are home to some of the largest trees in the world as well as the smallest animals. The multiple layers of canopy provide myriad resources and habitats for animals. At least 3 million species are thought to inhabit tropical forests although this number may actually be 10 or more times greater. Only about 500,000 species have actually been scientifically described, and many species are still to be discovered. Mammals, birds, reptiles, and amphibians, as well as insects and other invertebrates are more abundant here than any other place on Earth.

Tropical forests contribute far more to the world than just sustaining biodiversity. They provide habitat and homes to indigenous peoples, a multitude of natural products such as food, building materials, and medicines as well as ecosystem services such as soil stabilization and flood prevention, among others. These forests are important sources of carbon storage and play key roles in both regional and global climates.

Unfortunately, this great diversity of flora and fauna, forest products, and medicines is under tremendous threat. Tropical forests are disappearing at an alarming rate. Despite growing international concern, tropical forests continue to be destroyed at a pace exceeding 80,000 ac (32,000 ha) per day. Much of the remaining forests have been affected by human activities and no longer retain their original biodiversity. Clearing these forests affects biodiversity through habitat destruction, fragmentation of formally contiguous habitat, and increased edge effects. Deforestation of tropical forests has a global impact through species extinction, the loss of important ecosystem services and renewable resources, and the reduction of carbon sinks. This destruction, however, can be slowed, stopped, and in some cases even reversed.

Tropical forest loss is due to a complex combination of causes. Land use conversion for farming and ranching activities has led to the loss of forests, particularly in the Neotropics and Southeast Asia. Legal and illegal logging of rainforest trees through clear-cutting and selective cutting destroys forests and wildlife habitats, as well as changes the microclimate of the area. Fires in nearby cleared areas can spread to forests that are ill equipped to survive under these conditions. These changes influence surrounding forested areas, leading to negative feedback processes that further amplify the effects of local forest loss. Increases in population lead to increased need for land for settlements, as well as resources such as fuelwood and meat. Hunting contributes to the loss of many forest species, particularly large mammals.

Most of the tropical forests of the world lie in developing countries where population increases and economic pressures often outweigh the need to conserve the forest. Extraction of valuable minerals, gold, oil, and trees can provide temporary economic relief to these countries. The rates of extraction are unsustainable, however, and can lead to the total loss of forest and any potential revenue that an intact forest could have provided in the future.

As the world's tropical forests continue to disappear, considerable effort is needed to slow or stop these losses. International nongovernmental organizations such as the World Wildlife Fund, Conservation International, and the Rainforest Action Network are a few of the groups working on the problem. National and local governments and local community and conservation groups are involved in the preservation, protection, and sustainable harvest of these areas. As there are multiple factors causing the destruction of tropical forests, there must be multiple approaches to stop or lessen deforestation.

National and local parks and reserves can be successful if they are properly funded and if local communities are involved in the process of protecting local forests. Both large and small reserves can be important in supporting the diverse flora and fauna of the forests, smaller reserves can serve as corridors between larger areas allowing for the dispersal and migration of forest species. Reserves of this sort have been successful in Costa Rica, Bolivia, and Australia. In many areas, however, creating parks and reserves has neither stopped forest clearing by illegal loggers and developers nor improved the quality of living or economic opportunities for rural poor. Civil unrest and corruption has only worsened the situation. The problem with the traditional park approach in developing countries is that it fails to generate sufficient economic incentives for protecting the forest. Rainforests will only survive if they can be shown to provide economic benefits. To justify the costs of maintaining parks, local people and governments must see economic benefits in forgoing revenue from economic activities within protected areas.

Incorporating local communities into the planning and protection of the forest can increase the likelihood of success and provide needed income to the community. Low-impact ecotourism into these areas can increase economic opportunities while preserving the forest. The Quechau/Tacana people of the La Paz Department of Bolivia live in and around the Madidi National Forest. This forest is one of the most diverse intact rainforests in the entire Amazon. With the support of Conservation International and the Bolivian government, a lodge was constructed on the outskirts of the park. It is operated exclusively by the Quechua/Tacana people of the forest. The lodge provides employment and income for the local community. Their knowledge of the flora and fauna of the forests are valuable skills in leading small-scale tours through parts of the park. The majority of the community can maintain their traditional lifestyle supported by ecotourism income. The arrangement has been beneficial to the Quechua/Tacana people and the forest. Ecotourism in other areas can help fund conservation efforts by charging park entrance fees and employing locals as guides and as support staff in the handicraft and service sectors (hotels, restaurants, drivers, boat drivers, porters, cooks).

Creating sustainable plans to manage timber and mineral extraction in the forest can allow developing nations to grow their economies without destroying their means for future income. Most tropical countries rely heavily on the revenue generated by timber exports. With the demand for tropical woods increasing, these countries see short-term economic growth as a way to improve the quality of life for the country. However, this short-term rapid exploitation is unsustainable. Proper management of tropical forests can lead to a sustainable source for timber. Smaller and less-intensive timber processes, such as selective cutting rather than clear-cutting can provide tropical wood while retaining sufficient trees to grow and be removed later. Selective and careful extraction processes can cause less damage to the forest floor and understory where the next generation of trees are growing. Several Asian countries, such as Myanmar and Malaysia, have been somewhat successful in this endeavor. The system breaks down when political decisions, not based on ecological principles, are used to guide forestry practices.

An intact tropical forest provides a plethora of useful foods, materials, and medicines that can be sustainably harvested. Small-scale businesses selling goods produced from the forest can support local communities. Microcredit facilities can provide significant economic benefits to the local economy, promote entrepreneurship among local people, and provide sustainable economic activities without destroying the forests. Rainforest countries can also earn revenue by allowing scientists to develop products from native plant and animal species. One such example is Merck Pharmaceuticals. So that the company could look for plants with potential pharmaceutical applications, Merck made an agreement with Costa Rica that a portion of the proceeds from commercial compounds derived from its forests would go to the Costa Rican government. The Costa Rican government has guaranteed that some of the royalties will be set aside for conservation projects. Other companies have made similar agreements with tropical countries.

Concerns about global climate change and the increasing carbon released into the atmosphere by industrial nations has brought attention to the possibility of countries retaining their forests and receiving carbon credits, in the form of payments from developed countries looking to offset their carbon emissions. Carbon-offset programs are popular and can provide a way to motivate wealthy countries to pay for the benefits provided by forest conservation beyond their national borders.

Other steps can be taken to conserve the limited tropical forests that remain. Expanding protected areas and increasing surveillance of and patrols in protected areas using local communities can lead to more successful parks and reserves. Creating research facilities for training local scientists and guides allows the rainforest country to build its intellectual capital to grow its economy and make the best use of the country's resources. Establishing programs that promote sustainable use can increase the standard of living for people living around protected areas and help to ensure that the forests will remain.

The Tropical Forest Biomes of the world are ancient, unique, and valuable places. They contain rich diversity and provide abundant resources and services for the planet. However, tropical forests are quickly disappearing. As their destruction was initiated by humans, so must be the solutions to stop their destruction. A concerted effort by people in developed and developing nations working together is necessary to retain and protect these important biomes and the benefits they provide. Each one of us can help to save these forests by our actions and efforts. By understanding more about the ecology and biogeography of these biomes, you are one step closer to being part of the solution.

Further Readings

Books

Butler, Rhett. 2007. *How to Save Rainforests*. Mongabay.com. http://rainforests.mongabay.com.

Kallen, S. A., ed. 2006. *Rainforests: At Issue*. San Francisco: Thomson Gale.

Miller, K., and Laura Tangley. 1991. *Trees of Life: Saving Tropical Forests and Their Biological Wealth*. Boston: Beacon Press.

Newman, A. 2002. *Tropical Rainforest*. New York: Checkmark Books.

World Resources International. 2005. *World Resources 2005: The Wealth of the Poor*. Washington, DC: World Resources Institute.

Internet Source

Rainforest Action Network. 2007. *About Rainforests*. Rainforest Action Network. http://www.ran.org.

Glossary

Adaptation. Any inherited aspect of morphology, physiology, or behavior that improves a species chances of long-term survival and reproductive success in a particular environment.

Annual. A plant that completes its life cycle from seed to mature reproducing individual to death in a year's time or less.

Artiodactyl. Any of the even-toed ungulates, including antelopes, bison, and giraffes. With digestive systems geared to the fermentation of grasses, these mammals came to dominate the world's grasslands.

Biodiversity. The total variation and variability of life in a given region. Determined at various scales, including genes, species, and ecosystems.

Biome. A large subcontinental region of similar vegetation, animal life, and soils adapted to the area's physical environmental conditions such as climate or some disturbance factor.

Browser. A plant-eating animal that specializes on leaves, stems, or twigs. Compare with grazer.

Bulrush. A grass-like plant (graminoid) in the genus *Scirpus*. Associated with wetlands, it bears clusters of small brown florets.

C_3 Grass. A cool season grass that uses the chemical pathways of the Calvin Cycle to assimilate carbon dioxide into 3-carbon molecules during photosynthesis. Most green plants fall into this category.

C_4 Grass. A warm season grass that uses 4-carbon molecules to convert carbon dioxide to organic molecules during photosynthesis. Common among tropical grasses.

Cape Floristic Province. Another term for the South African Floristic Kingdom, a small area surrounding the Cape of Good Hope and renowned for its extremely high plant species diversity.

Caviomorph (rodent). One of a group of distinctly South American rodents that includes the cavies, capybara, and mara. The porcupine is the only caviomorph in North America.

Climate. The general weather patterns expected during an average year. The main factors are temperature and precipitation. See also **weather.**

Cold season grass. Grasses that grow best at moderate temperatures and often become dormant during the hottest part of the year, resuming growth begun in early spring when the lower temperatures of autumn arrive. These are C_3 grasses typical of the temperate grasslands.

Culm. The stem of a grass or other graminoid.

Cushion plant. A low, many-stemmed plant that grows as a dense mound.

Disturbance. An event that disrupts an ecosystem and damages or destroys some part of it. Origin may be biological (such as overgrazing) or physical (such as edaphic conditions, fire, or flood).

Domesticated. Produced by the selective breeding of humans or by natural adaptation to human-dominated environments.

Edaphic. Pertaining to conditions of the substrate or soil. Those that affect plant growth include nutrient-depletion, poor drainage, excessive drainage, and presence of a hardpan.

El Niño. A seasonal weather phenomenon that affects the equatorial Pacific, especially off the west coast of South America. During these events of December, normal high-pressure systems that make the coast exceptionally dry are replaced by low pressure, high humidity, and even rain. Severe, prolonged El Niños can affect weather patterns around the world.

Endemic. Native to and restricted to a particular geographic area.

Exotic. Nonnative, introduced, alien.

Fauna. All the animal species in a given area, or some subset of them such as the bird fauna or grazing fauna.

Forb. An herb with broad leaves and soft, nonwoody stem. Wildflowers are typical of this growthform.

Fossorial. Burrowing. Adapted to living underground

Fynbos. The local name for mediterranean scrub vegetation in South Africa's Cape Floristic Province.

Graminoid. The growthform of grasses, sedges, rushes, and reeds. A type of herb.

Grass. Any flowering plant of the family Poaceae (also known as the family Graminae).

Grazer. A plant-eating animal that consumes primarily grasses. Compare to browser.

Hardpan. A dense rock-like layer of substrate that is difficult for water or roots to penetrate.

Herb. A nonwoody plant that dies down each year. May be an annual or a perennial. This growthform includes graminoids and forbs.

Herbaceous. Having the characteristics of an herb.

Honeydew. A sweet, sticky substance secreted by aphids and some scale insects.

Hotspot. An area of high biodiversity, usually threatened and requiring protection.

Humus. Well-decayed plant matter that is collidal in size and assumes a dark-brown color. Since it helps hold moisture and nutrients in a soil, humus content is an indicator of soil fertility.

Ion. A particle bearing a negative or positive charge.

ITCZ (Intertropical Convergence Zone). The contact zone between the Trade Winds of the Northern and Southern hemispheres. Shifts its position north and south of the Equator with the seasons and, when overhead, usually brings rain.

Lagomorph. A rabbit, hare, or pika. A member of the Order Lagamorpha.

Laterite. A hardpan that forms in red tropical soils (oxisols) due to the concentration of iron oxide.

Latitude. The distance of a point north or south of the Equator (0° latitude), measured in degrees.

Leaching. The process in which dissolved substances are removed from the substrate by the downward percolation of water.

Legume. Any plant that is a member of the pea and bean family (Leguminoseae) or, in other taxonomic schemes, Mimosaceae, Caesalpinaceae, or Fabaceae. Many of these plants have symbiotic relationships with rhizobial bacteria on their roots that fix atmospheric nitrogen into nitrates, which are important plant nutrients.

Loess. Powder-like material deposited by the wind. The most fertile temperate soils, the chernozems, are generally developed on this type of substrate.

Marsupial. A nonplacental mammal of the order Marsupalia whose young develop attached to teats in a pouch. Considered an early or primitive form of mammal, marsupials dominate the mammalian fauna of Australia and are also diverse in the Neotropics.

Mediterranean. Refers to regions or climate patterns in which winter is the rainy season and summers are dry.

Migration. The seasonal movement of a population from one area to another with a return to the original range when seasons change.

Monsoon. A wind that reverses its direction seasonally. An onshore flow typifies the warm season and an offshore flow occurs during the cold season. The Asian monsoon is most powerful and dominates the climate of the vast Indian Ocean region. More localized monsoonal systems occur elsewhere, as in the American Southwest.

Neotropical. Pertaining to the region from southern Mexico and the Caribbean to southern South America and to animals and plants restricted or nearly restricted to this region.

Nitrogen-fixing. The ability to convert elemental or pure nitrogen to nitrates. Only certain microorganisms can do this and hence they are vital to higher lifeforms that require nitrogen but cannot utilize it in its pure form. Rhizobial bacteria and cyanobacteria in the soil are important nitrogen-fixers in terrestrial ecosystems.

Perennial. A plant that lives two or more years.

Perissodactyl. Any of the odd-toed ungulates. Living forms include zebras, rhinoceroses, and tapirs.

Polar Front. In the global atmospheric circulation system, the contact zone between cold polar air and warmer subtropical airmasses. Associated with uplift at the subpolar lows and often stormy, wet weather patterns in the mid-latitudes.

Rainshadow. A dry region that develops on the lee (downwind) side of mountain ranges. Low amounts of precipitation are the result of the warming of airmasses as they descend the mountain slopes.

Ratite. A flightless bird lacking a broad sternum for the attachment of wing muscles. Large tropical grassland birds well adapted to running, ratites include ostriches, emus, cassowaries, and rheas.

Reed. A large hollow-stemmed grass of genera such as *Arundo* and *Phragmites*, possessing plume-like inflorescences.

Rhizome. Underground stems from which new plantlets or tillers arise, leading to the sod-forming habit of some grasses.

Ruminant. An artiodactyl that possesses a four-chambered stomach in which grasses ferment and that chews its cud.

Sedge. A grass-like flowering plant in the family Cypercaceae. Members of the genus *Carex* usually prefer damper conditions than grasses.

Semiarid. Referring to climatic conditions in which there is too little precipitation to support forests, but no so little that deserts prevail. In the mid-latitudes, semiarid regions usually receive between 10 and 20 in of precipitation a year and support natural grasslands.

Stolon. A horizontal stem that forms at ground level and gives rise to new tillers in grasses.

Stoma or stomate. A tiny pore in the outer layer of a leaf through which gases, including water vapor, are exchanged with the atmosphere. The plural is stomata.

Subshrub. A hard-stemmed shrub in which the upper branches die back during the non-growing season. In a sense, a shrub that acts like a perennial forb.

Succulent. A growthform that permits the storage of water in some of its tissues. Plants may be leaf succulents, stem succulents, or have special underground organs for storing water.

Tiller. A daughter plant or new plantlet forming from a grass's stolons or rhizomes.

Trade Winds. The strong, constant easterly winds of tropical latitudes.

Tropics. The latitudinal zone on Earth that lies between 23°30′ N and 23°30′ S, that is, between the Tropic of Cancer and the Tropic of Capricorn.

Tussock. A growthform of grasses and sedges in which individuals grow in clumps, forming visible hummocks.

Ungulate. A hoofed mammal.

Vegetation. The general plant cover of an area described in terms of its structure and appearance and not the species that comprise it.

Warm Season Grass. A grass that photosynthesizes and grows best under high temperature conditions. These are C_4 grasses found in the tropics and in temperate grasslands with hot summers.

Weather. The state of the atmosphere at any given moment. Includes atmospheric pressure, temperature, humidity, and type of precipitation (if any).

Weathered. Pertaining to bedrock that has undergone physical and/or chemical breakdown into small particles, even ions.

Weed. A plant adapted to invade disturbed sites. Generally short-lived, they are good dispersers and fast growers.

Bibliography

General Introduction to Tropical Forests

Achard, F., H. D. Eva, H. J. Stibig, P. Mayaux, J. Gallego, T. Richards, and J. P. Malingreau. 2002. "Determination of Deforestation Rates of the World's Humid Tropical Forests." *Science* 297: 999–1002.

Allee, W. C., and Karl P. Schmidt. 1951. *Ecological Animal Geography*. New York: John Wiley & Sons, Inc.

Aubert De La Rue, E., Francois Bourliere, and Jean-Paul Harroy. 1957. *The Tropics*. New York: Alfred A. Knopf.

Brown, James, and Arthur Gibson. 1983. *Biogeography*. St. Louis, MO: The C.V. Mosby Company.

Chape, S., S. Blyth, I. Fish, P. Fox, and M. Spalding. 2003. *United Nations List of Protected Areas*. Cambridge: International Union for Conservation of Nature.

Cool Planet. 2002. "Tropical Rain Forest." Oxfam. http://www.oxfam.org.uk/coolplanet/ontheline/explore/nature/trfindex.htm.

Dale, V. H., and M. A. Pedlowski. 1992. "Farming the Forests." *Forum for the Applied Research and Public Policy* 7: 20–21.

DeFries, Ruth S., R. A. Houghton, M. C. Hansen, C. B. Field, D. Skole, and J. Townshend. 2002. "Carbon Emissions from Tropical Deforestation and Regrowth Based on Satellite Observations for the 1980s and 1990s." *Proceedings of the National Academy of Science* 99: 14256–14261.

Denslow, J., and C. Padoch. 1988. *People of the Tropical Rainforest*. Berkeley, CA: Unversity of California Press.

Ehrlich, Paul R. 1988. "The Loss of Diversity: Causes and Consequences." In *Biodiversity*, ed. E. O. Wilson and F. M. Peter, 21–27. Washington, DC: National Academy Press.

Emmons, L. 1990. *Neotropical Rainforest Mammals.* Chicago: University of Chicago Press.

FAO (Food and Agriculture Organization of the United Nations). 1993. *Forest Resources Assessment 1990: Tropical Countries.* Vol. 112. Rome: FAO.

Jenny, Hans. 1983. *The Soil Resource.* New York: Springer-Verlag.

Kellman, Martin, and Rosanne Tackaberry. 1997. *Tropical Environments: The Functioning and Management of Tropical Ecosystems.* London: Routledge.

Lai, R., J. M. Kimble, and B. A. Stewart, eds. 2000. *Global Climate Change and Tropical Ecosystems: Advances in Soil Science.* London: CRC Press.

Laurance, W. F. 1999. "Reflections on the Tropical Deforestation Crisis." *Biological Conservation* 91: 109–117.

Laurance, W. F., and R. O. Beirregaard, eds. 1997. *Tropical Forest Remnants: Ecology, Management and Conservation of Fragmented Communities.* Chicago: University of Chicago Press.

Lieth, H., and M. J. A. Wegener, eds. 1989. *Tropical Rain Forests Ecosystems: Biogeographical and Ecological Studies.* New York: Elsevier.

Mabberley, D. J. 1991. *Tropical Rain Forest Ecology.* Oxford: Chapman and Hall.

Marcon, E., and Manuel Mongini. 1984. *The World Encyclopedia of Animals.* London: Orbis Publishers.

Miller, K., and Laura Tangley. 1991. *Trees of Life: Saving Tropical Forests and Their Biological Wealth.* Boston: Beacon Press.

Missouri Botanical Garden VAST. 2007. "Web TROPICOS." Missouri Botanical Garden. http://mobot.mobot.org/W3T/Search/vast.html.

Myers, N. 1980. *Conversion of Tropical Moist Forests.* Washington, DC: National Academy of Sciences.

Myers, N. 1990. "The World's Forest and Human Populations: The Environmental Interconnections." *Population and Development Review* 16: 237–251.

Myers, N. 1994. "Tropical Deforestation: Rates and Patterns. The Causes of Tropical Deforestation." In *The Economic and Statistical Analysis of Factors Giving Rise to the Loss of the Tropical Forests,* ed. K. Brown and D. W. Pierce, 27–40. Vancouver: University of British Columbia Press.

Myers, N. 1997. "The World's Forests and their Ecosystem Services." In *Nature's Services: Societal Dependence on Natural Ecosystems,* ed. G.C. Daily, 215–235. Washington, DC: Island Press.

Myers, N., and A. H. Knoll. 2001. "The Biotic Crisis and the Future of Evolution." *Proceedings of the National Academy of Sciences* 98: 5389–5392.

Myers, N., and T. Pages. 1988. "Tropical Forests and Their Species: Going, Going …?" In *Biodiversity,* ed. E. O. Wilson and M. Peters, 28–35. Washington, DC: National Academy Press.

Myers, P. 2001. "Animalia: Animal Diversity Web." University of Michigan Museum of Zoology. http://animaldiversity.ummz.umich.edu/site/accounts/information/Animalia.html.

National Geographic and World Wildlife Fund. 2001. "Wild World, Terrestrial Ecoregions of the World." http://www.nationalgeographic.com/wildworld/terrestrial.html.

National Science Foundation. 2006. "Tropical Rainforest Nutrients Linked to Global Carbon Dioxide Levels." National Science Foundation. http://www.nsf.gov/news/news_summ.jsp?cntn_id=107047&org=NSF.

Nowak, R. 1999. *Walker's Mammals of the World*. Vols. 1 and 2. Baltimore: Johns Hopkins University Press.

Nowak, R. 2005. *Walker's Carnivores of the World*. Baltimore: Johns Hopkins University Press.

Olson, D. M., E. Dinerstein, E. D. Wikramanayake, N. D. Burgess, G. V. N. Powell, E. C. Underwood, J. A. D'Amico, I. Itoua, H. E. Strand, J. C. Morrison, C. J. Loucks, T. F. Allnutt, T. H. Ricketts, Y. Kura, J. F. Lamoreux, W. W. Wettengel, P. Hedao, and K. R. Kassem. 2001. "Terrestrial Ecoregions of the Worlds: A New Map of Life on Earth." *Bioscience* 51: 933–938.

Peters, W. J., and L. F. Neuenschwander. 1988. *Slash and Burn: Farming in the Third World Forest*. Moscow: University of Idaho Press.

Pimm, S. L., G. J. Russell, J. L. Gittleman, and T. M. Brooks 1995. "The Future of Biodiversity." *Science* 269: 347–350.

Raven, P. H. 1988. "Our Diminishing Tropical Forests." In *Biodiversity*, ed. E. O. Wilson and F. M. Peter, 119–122. Washington, DC: National Academy Press.

Richards, P. W. 1996. *The Tropical Rain Forest: An Ecological Study*. Cambridge: Cambridge University Press.

Sutton, Ann, and Myron Sutton. 1979. *Wildlife of the Forests*. New York: Chanticleer Press, Inc.

United Nations Population Division. 2005. "World Population Prospects: The 2004 Revision Highlights." Department of Economic and Social Affairs, United Nations. http://www.un.org/esa/population/publications/WPP2004/2004Highlights_finalrevised.pdf.

USDA-ARS (U.S. Department of Agriculture, Agricultural Research Service). National Genetic Resources Program. 2007. "Germplasm Resources Information Network GRIN." [Online Database]. National Germplasm Resources Laboratory, Beltsville, MD. http://www.ars-grin.gov/cgi-bin/npgs/html/taxon.pl?101938.

USDA-NRCS (United States Department of Agriculture, Natural Resources Conservation Service). 2007. The PLANTS Database. National Plant Data Center, Baton Rouge, LA. http://plants.usda.gov.

White, R. P., S. Murray, and M. Rohweder. 2000. *Pilot Analysis of Global Ecosystems*. Washington, DC: World Resources Institute.

Wilson, Don E., and F. Russell Cole. 2000. *Common Names of Mammals of the World*. Washington DC: Smithsonian Institution Press.

Wilson, E. O. 1992. *The Diversity of Life*. Cambridge, MA: Harvard University Press.

Wilson, E. O., and F. M. Peter, eds. 1988. *Biodiversity*. Washington, DC: National Academy Press.

Woodward, Susan. 2003. *Biomes of Earth*. Westport, CT: Greenwood Publishing Group.

Woodward, Susan. 2004. "Introduction to Biomes." Radford University, Department of Geography. http://www.radford.edu/~swoodwar/CLASSES/GEOG235/biomes/intro.html.

Tropical Rainforests (General)

Adams, J. 1994. "The Distribution and Variety of Equatorial Rain Forests." Oak Ridge National Laboratory. http://www.esd.ornl.gov/projects/qen/rainfo.html.

Allen, J. C., and D. F. Barnes. 1985. "The Causes of Deforestation in Developing-Countries." *Annals of the Association of American Geographers* 75: 163–184.

Bermingham, Eldredge, Christopher W. Dick, and Craig Moritz, eds. 2005. *Tropical Rainforests: Past, Present, and Future.* Chicago: University of Chicago Press.

Boumans, R., R. Costanza, J. Farley, M. A. Wilson, R. Portela, J. Rotmans, F. Villa, and M. Grasso. 2002. "Modeling the Dynamics of the Integrated Earth System and the Value of Global Ecosystem Services Using the GUMBO Model." *Ecological Economics* 41: 529–560.

Butler, R. 2007. "Rainforests." Mongabay.com. http://rainforests.mongabay.com/.

Chapin, F. S., E. S. Zavaleta, V. T. Eviner, R. L. Naylor, P. M. Vitousek, H. L. Reynolds, D. U. Hooper, S. Lavorel, O. E. Sala, S. E. Hobbie, M. C. Mack, and S. Díaz. 2000. "Consequences of Changing Biodiversity." *Nature* 405 (6783): 234–242.

Collins, M., ed. 1990. *The Last Rain Forests.* New York: Oxford University Press.

Conklin, H. C. 1961. "The Study of Shifting Cultivation." *Current Anthropology* 2 (1): 27–61.

Costanza, Robert, Ralph d'Arge, Rudolf de Groot, S. Farber, M. Grasso, B. Hannon, K. Limburg, S. Naeem, R. V. O'Neill, J. Paruelo, R. G. Raskin, P. Sutton, and M. van den Belt. 1997. "The Value of the World's Ecosystem Services and Natural Capital." *Nature* 387: 253–260.

Daily, G. C., ed. 1997. *Nature's Services: Societal Dependence on Natural Ecosystems.* Washington, DC: Island Press.

Daily, G. C., and Paul Ehrlich. 1992. "Population, Sustainability, and Earth's Carrying Capacity: A Framework for Estimating Population Sizes and Lifestyles that Could be Sustained Without Undermining Future Generations." *BioScience* 42 (10): 761–771.

Davis, Charles., C. Campbell, O. Webb, Kenneth J. Wurdack, Carlos A. Jaramillo, and Michael J. Donoghue. 2005. "Explosive Radiation of Malpighiales Supports a Mid-Cretaceous Origin of Modern Tropical Rain Forests." *The American Naturalist* 165 (3): 36–65.

DeFries, Ruth S., R. A. Houghton, M. C. Hansen, C. B. Field, D. Skole, and J. Townshend. 2002. "Carbon Emissions from Tropical Deforestation and Regrowth Based on Satellite Observations for the 1980s and 1990s." *Proceedings of the National Academy of Science* 99: 14256–14261.

Ehrlich, Paul R. 1988. "The Loss of Diversity: Causes and Consequences." In *Biodiversity,* ed. E. O. Wilson and F. M. Peter, 21–27. Washington, DC: National Academy Press.

Gaston, K. J. 2000. "Global Patterns in Biodiversity." *Nature* 405: 220–227.

Gay, K. 2001. *Rainforests of the World. A Reference Handbook.* Santa Barbara, CA: ABC-CLIO, Inc.

Golley, F. B., ed. 1983. *Tropical Rain Forest Ecosystems.* Amsterdam: Elsevier Scientific Publishing Company.

Graydon, N. 2006. "Rainforest Action Network: The Inspiring Group Bringing Corporate America to its Senses." *The Ecologist* February 36 (1): 40–47.

Hannah, L., D. Lohse, C. Hutchinson, J. L. Carr, and A. Lankerani. 1994. "A Preliminary Inventory of Human Disturbance on World Ecosystems." *Amibo* 23: 246–250.

Hoekstra, J. M., T. M. Boucher, T. H. Ricketts, and C. Roberts. 2005. "Confronting a Biome Crisis: Global Disparities of Habitat Loss and Protection." *Ecology Letters* 8: 23–29.

Holdridge, L. R. 1967. *Life Zone Ecology.* San Jose, Costa Rica: Tropical Science Center.

Holm-Nielsen, L. B., I. C. Nielsen, and H. Balslev. 1988. *Tropical Rainforests: Botanical Dynamics, Speciation and Diversity.* London: Academic Press.

Jordan, C. F. 1985. *Nutrient Cycling in Tropical Forest Ecosystems*. New York: John Wiley & Sons, Inc.

Kallen, S. A., ed. 2006. *Rainforests: At Issue*. San Francisco: Thomson Gale.

Lambertini, M. 1992. *A Naturalist's Guide to the Tropics*. Chicago: University of Chicago Press.

Laurance, W. F. 1999. "Reflections on the Tropical Deforestation Crisis." *Biological Conservation* 91: 109–117.

Laurance, W. F., and R. O. Beirregaard, eds. 1997. *Tropical Forest Remnants: Ecology, Management and Conservation of Fragmented Communities*. Chicago: University of Chicago Press.

Morely, R. J. 2000. *Origin and Evolution of Tropical Rainforests*. West Sussex, UK: John Wiley & Sons, Ltd.

Newman, A. 2002. *Tropical Rainforest*. New York: Checkmark Books.

Nowak, R. 1999. *Walker's Mammals of the World*. Vols. 1 and 2. Baltimore: Johns Hopkins University Press.

Nowak, R. 2005. *Walker's Carnivores of the World*. Baltimore: Johns Hopkins University Press.

Park, Chris. 1992. *Tropical Rainforests*. New York: Routledge Publishing.

Prance, G. T., ed. 1982. *Biological Diversification in the Tropics*. New York: Columbia University Press.

Primack, R., and R. Corlett. 2005. *Tropical Rain Forests*. Oxford: Blackwell Publishing.

Sanchez, P. A. 1989. "Soils. Tropical Rain Forests Ecosystems." In *Biogeographical and Ecological Studies,* ed. H. Lieth and M. J. A. Werger, 73–87. New York: Elsevier.

Terbough, J. 1993. *Diversity and the Tropical Rain Forest*. New York: Scientific American Library.

Tosi, J., and R. F. Voertman. 1964. "Some Environmental Factors in the Economic Development of the Tropics." *Economic Geography* 40: 189–205.

Vandermeer, J., and Ivette Perfecto. 1995. *Breakfast of Biodiversity: The Truth about Rain Forest Destruction*. Oakland, CA: The Institute for Food and Development Policy.

Vitousek, P. M., H. A. Mooney, J. Lubchenco, and J. M. Melillo. 1997. "Human Domination of Earth's Ecosystems." *Science* 277 (5325): 494–499.

Tropical Seasonal Forests (General)

Bullock, S. H., H. A. Mooney, and E. Medina, eds. 1995. *Seasonally Dry Tropical Forests.* Cambridge: Cambridge University Press.

Cochrane, M. A. 2003. "Fire Science for Rainforests." *Nature* 421 (6926): 913–919.

Cochrane, M. A., and M. D. Schulze. 1999. "Fire as a Recurrent Event in Tropical Forests of the Eastern Amazon: Effects of Forest Structure, Biomass and Species Composition." *Biotropica* 31: 2–16.

Conklin, H. C. 1961. "The Study of Shifting Cultivation." *Current Anthropology* 2 (1): 27–61.

Ewel, J. J. 1977. "Differences between Wet and Dry Forest Successional Tropical Ecosystems." *International Journal of Tropical Ecology and Geography* 1: 103–117.

Janzen, H. D. 1988. "Tropical Dry Forest: The Most Endangered Major Tropical Ecosystem." In *Biodiversity,* ed. E. O. Wilson and F. M. Peter, 130–137. Washington, DC: National Academy Press.

Lieberman, D., and M. Lieberman. 1984. "The Causes and Consequences of Synchronous Flushing in a Dry Tropical Forest." *Biotropica* 16: 193–201.

Miles, L., A. C. Newton, R. S. DeFries, C. Ravilious, I. May, S. Blyth, V. Kapos, and J. E. Gordon. 2006. "A Global Overview of the Conservation Status of Tropical Dry Forests." *Journal of Biogeography* 33: 491–505.

Missouri Botanical Garden VAST. 2007. "Web TROPICOS." Missouri Botanical Garden. http://mobot.mobot.org/W3T/Search/vast.html.

Morely, R. J. 2000. *Origin and Evolution of Tropical Rainforests.* West Sussex, UK: John Wiley & Sons, Ltd.

Murphy, P. G., and Ariel E. Lugo. 1986. "Ecology of Tropical Dry Forest." *Annual Review of Ecology and Systematics* 17: 67–88.

Myers, P. 2001. "Animalia: Animal Diversity Web." University of Michigan Museum of Zoology. http://animaldiversity.ummz.umich.edu/site/accounts/information/Animalia .html.

National Geographic and World Wildlife Fund. 2001. "Wild World, Terrestrial Ecoregions of the World." http://www.nationalgeographic.com/wildworld/terrestrial.html.

Nowak, R. 1999. *Walker's Mammals of the World.* Vols. 1 and 2. Baltimore: Johns Hopkins University Press.

Nowak, R. 2005. *Walker's Carnivores of the World.* Baltimore: Johns Hopkins University Press.

Peters, W. J., and L. F. Neuenschwander. 1988. *Slash and Burn: Farming in the Third World Forest.* Moscow: University of Idaho Press.

Prance, G. T., ed. 1982. *Biological Diversification in the Tropics.* New York: Columbia University Press.

USDA-ARS (United States Department of Agriculture, Agricultural Research Service). National Genetic Resources Program. 2007. "Germplasm Resources Information Network GRIN." [Online Database]. National Germplasm Resources Laboratory, Beltsville, MD. http://www.ars-grin.gov/cgi-bin/npgs/html/taxon.pl?101938.

Walter, H. 1971. *Ecology of Tropical and Subtropical Vegetation.* London: Oliver & Boyd.

Wilson, Don E., and F. Russell Cole. 2000. *Common Names of Mammals of the World.* Washington DC: Smithsonian Institution Press.

Woodward, Susan. 2003. *Biomes of Earth.* Westport, CT: Greenwood Publishing Group.

Regional Expressions of the Tropical Rainforest and Seasonal Forests

Neotropics

Arvigo, Rosita, and Michael Balick. 1993. *Rainforest Remedies: One Hundred Healing Herbs of Belize.* Twin Lakes, WI: Lotus Press.

Ceballos, G. 1995. "Vertebrate Diversity, Ecology and Conservation in Neotropical Dry Forests." In *Seasonally Dry Tropical Forests,* ed. S. H. Bullock, H. A. Mooney, and E. Medina, 195–220. Cambridge: Cambridge University Press.

Cochrane, M. A., and M. D. Schulze. 1999. "Fire as a Recurrent Event in Tropical Forests of the Eastern Amazon: Effects of Forest Structure, Biomass and Species Composition." *Biotropica* 31: 2–16.

Emmons, L. 1990. *Neotropical Rainforest Mammals.* Chicago: University of Chicago Press.

Fearnside, P. M. 1986. *Human Carrying Capacity of the Brazilian Rain Forest.* New York: Columbia University Press.

Fearnside, P. M. 1993. "Deforestation in Brazilian Amazon: The Effect of Population and Land Tenure." *Ambio* 8: 537–545.

Fearnside, P. M. 2001. "Soybean Cultivation as a Threat as a Threat to the Environment in Brazil." *Environmental Conservation* 28: 23–38.

Garrigues, Richard, and Robert Dean. 2007. *The Birds of Costa Rica: A Field Guide.* Ithaca, NY: Cornell University Press.

Gentry, A. H. 1982. "Neotropical Floristic Diversity: Phytogeographical Connections Between Central and South America, Pleistocene Climatic Fluctuations, or Accident of the Andean Orogeny." *Annals of the Missouri Botanical Garden* 69: 557–593.

Gentry, A. H., ed. 1990. *Four Tropical Rainforests.* New Haven, CT: Yale University Press.

George, W., and Rene Lavocat. 1993. *The Africa-South America Connection.* Oxford: Clarendon Press.

Hecht, S. B., and A. Cockburn. 1989. *The Fate of the Forest: Developers, Destroyers, and Defenders of the Amazon.* New York: Verso.

Hoekstra, J. M., T. M. Boucher, T. H. Ricketts, and C. Roberts. 2005. "Confronting a Biome Crisis: Global Disparities of Habitat Loss and Protection." *Ecology Letters* 8: 23–29.

Hogue, C. 1993. *Latin American Insects and Entomology.* Berkeley: University of California Press.

Holbrook, N. M., J. L. Whitbeck, and H. A. Mooney. 1995. "Drought Responses of Neotropical Dry Forest Trees." In *Seasonally Dry Tropical Forests,* ed. S. H. Bullock, H. A. Mooney, and E. Medina, 243–276. Cambridge: Cambridge University Press.

Holdridge, L. R. 1967. *Life Zone Ecology.* San Jose, Costa Rica: Tropical Science Center.

Holm-Nielsen, L. B., I. C. Nielsen, and H. Balslev. 1988. *Tropical Rainforests: Botanical Dynamics, Speciation and Diversity.* London: Academic Press.

Janzen, H. D. 1988. "Tropical Dry Forest: The Most Endangered Major Tropical Ecosystem." In *Biodiversity,* ed. E. O. Wilson and F. M. Peter, 130–137. Washington, DC: National Academy Press.

Kapos, V. 1989. "Effects of Isolation on the Water Status of Forest Patches in the Brazilian Amazon." *Journal of Tropical Ecology* 5: 173–185.

Kaspari, M., and Sean O'Donnell. 2003. "High Rates of Army Ant Raids in the Neotropics and Implications for Ant Colony and Community Structure." *Evolutionary Ecology Research* 5: 933–939.

Kircher, John. 1997. *A Neotropial Companion.* Princeton, NJ: Princeton University Press.

Laurance, W. F., T. E. Lovejoy, H. L. Vasconcelos, E. M. Bruna, R. K. Didham, P. C. Stouffer, C. Gascon, R. O. Bierregaard, S. G. Laurance, and E. Sampaio. 2002. "Ecosystem Decay of Amazonian Forest Fragments: A 22-Year Investigation." *Conservation Biology* 16: 605–618.

Leigh, E. G. 1999. *Tropical Forest Ecology: A View from Barro Colorado Island.* New York: Oxford University Press.

Malcolm, J. R. 1994. "Edge Effects in Central Amazonian Forest Fragments." *Ecology* 75: 2438–2445.

Martinez-Yrizar, A. Burquez, and M. Maass. 2000. "Structure and Functioning of a Tropical Deciduous Forest in Western Mexico." In *The Tropical Deciduous Forest of Alamos: Biodiversity of a Threatened Ecosystem in Mexico,* ed. R. H. Robichaux and D. A. Yetman, 19–35. Tucson: University of Arizona Press.

Medina, E. 1995. "Diversity of Life Forms of Higher Tropical Plants in Neotropical Dry Forests." In *Seasonally Dry Tropical Forests,* ed. S. H. Bullock, H. A. Mooney, and E. Medina, 221–243. Cambridge: Cambridge University Press.

Meggers, B. J., Edward S. Ayensu, and D. Duckworth. 1973. *Tropical Forest Ecosystems in Africa and South America: A Comparative Review.* Washington, DC: Smithsonian Institution Press.

Mieres, M. M., and Lee A. Fitzgerald. 2006. "Monitoring and Managing the Harvest of Tegu Lizard in Paraguay." *Journal of Wildlife Management* 70 (6): 1723–1734.

Montgomery, G. G. 1985. *The Evolution and Ecology of Armadillos, Sloths and Vermilinguas.* Washington, DC: Smithsonian Institution Press.

Robichaux, R. H., and D. A. Yetman, eds. 2000. *The Tropical Deciduous Forest of Alamos: Biodiversity of a Threatened Ecosystem in Mexico.* Tucson: University of Arizona Press.

Rodriguez Mata, Jorge, Francisco Erize, and Maurice Rumboll. 2006. *Birds of South America: Non-Passerines: Rheas to Woodpeckers.* Princeton Illustrated Checklists. Princeton, NJ: Princeton University Press.

Russell, S. M. 2000. "Birds of the Tropical Deciduous Forest of the Alamos, Sonora Area." In *The Tropical Deciduous Forest of Alamos: Biodiversity of a Threatened Ecosystem in Mexico,* ed. R. H. Robichaux and D. A. Yetman, 200–244. Tucson: University of Arizona Press.

Sarmiento, G. 1972. "Ecological and Floristic Convergences between Seasonal Plant Formations of Tropical and Subtropical South America." *The Journal of Ecology* 60 (2): 367–410.

Steininger, M. K., C. J. Tucker, P. Ersts, T. J. Killeen, Z. Villegas, and S. B. Hecht. 2001. "Clearance and Fragmentation of Tropical Deciduous Forest in the Tierras Bajas, Santa Cruz, Bolivia." *Conservation Biology* 15: 856–866.

Stotz, D., John W. Fitzpatrick, Theodore A. Parker III, and Debra A. Moskovits. 1996. *Neotropical Birds: Ecology and Conservation.* Chicago: University of Chicago Press.

Taylor, K. I. 1988. "Deforestation and Indians in Brazilian Amazonia." In *Biodiversity,* ed. E. O. Wilson and F. M. Peter, 138–144. Washington, DC: National Academy Press.

Uhl, C., and R. Buschbacher. 1985. "A Disturbing Synergism between Cattle Ranch Burning Practices and Selective Tree Harvesting in the Eastern Amazon." *Biotropica* 17 (4): 265–268.

Uhl, C., and J. B. Kaufman. 1990. "Deforestation, Fire Susceptibility, and Potential Tree Responses to Fire in the Eastern Amazon." *Ecology* 71: 437–449.

Viña, A., and J. Cavalier. 1999. "Deforestation Rates 1938–1988 of Tropical Lowland Forests on the Andean Foothills of Columbia." *Biotropica* 31: 31–36.

Wallace, Alfred R. 1972. *A Narrative of Travels on the Amazon and Rio Negro, 1889.* New York: Dover.

Zuchowski, Willow. 2007. *Tropical Plants of Costa Rica. A Guide to Native and Exotic Flora.* Ithaca, NY: Cornell University Press.

Africa and Madagascar

Bradt, Hilary. 1994. *Guide to Madagascar.* Old Saybrook, CT: The Globe Pequot Press.

Burgess, N., Jennifer D'Amico, Emma Underwood, Eric Dinerstein, David Olson, Illanga Itoua, Jan Schipper, Taylor Ricketts, and Kate Newman. 2004. *Terrestrial Ecoregions of Africa and Magascar: A Conservation Assessment.* Washington, DC: Island Press.

Delaney, M. J., and D. C. D. Happold. 1979. *Ecology of African Mammals.* London: Longman Group Limited.

Dominy, N. J., and B. Duncan. 2001. "GPS and GIS Methods in an African Rain Forest: Applications to Tropical Ecology and Conservation." *Conservation Ecology* 5 (2): 6. http://www.ecologyandsociety.org/vol5/iss2/art6/.

Drucker-Brown, Susan. 1995. "The Court and the Kola Nut: Wooing and Witnessing in Northern Ghana." *The Journal of the Royal Anthropological Institute* 1: 129–143.

Hopkins, B. 1966. *Forest and Savanna*. London: Heinemann Press.

Jarosz, L. 1993. "Defining and Explaining Tropical Deforestation: Shifting Cultivation and Population Growth in Colonial Madagascar: 1896–1940." *Economic Geography* 69: 366–379.

Leach, M. 1994. *Rainforest Relations: Gender and Resource Use Among the Mende of Gola, Sierra Leone*. Washington, DC: Smithsonian Institution Press.

Leache, A. D., Mark-Oliver Rodel, Charles W. Linkem, Raul E. Diaz, Annika Hillers, and Matthew K. Fujita. 2006. "Biodiversity in a Forest Land: Reptiles and Amphibian of the West African Togo Hills." *Amphibian and Reptile Conservation* 4 (1): 22–45.

Lieberman, D., Milton Lieberman, and Claude Martin. 1987. "Notes on Seeds in Elephant Dung from Bia National Park, Ghana." *Biotropica* 19 (4): 365–369.

Lovett, J. C., and Samuel K. Wasser, eds. 1993. *Biogeography and Ecology of the Rain Forests of Eastern Africa*. Cambridge: Cambridge University Press.

Martin, C. 1991. *The Rainforests of West Africa: Ecology, Threats and Conservation*. Basel: Birkhäuser Verlag.

Menaut, J. C., M. LePage, and L. Abbadie. 1995. "Savannas, Woodlands and Dry Forests in Africa." In *Seasonally Dry Tropical Forests,* ed. S. H. Bullock, H.A. Mooney, and E. Medina, 64–93. Cambridge: Cambridge University Press.

Mott, M. 2004. "Bees, Giant African Rats Used to Sniff Landmines." *National Geographic News.* http://news.nationalgeographic.com/news/2004/02/0210_040210_minerats.html.

Nielsen, M. S. 1965. *Introduction to the Flowering Plants of West Africa*. London: University of London Press Ltd.

Owen, D. F. 1971. *Tropical Butterflies: The Ecology and Behaviour of Butterflies in the Tropics with Special Reference to African Species*. Oxford: Clarendon Press.

Pochron, Sharon, and Patricia C. Wright. 2005. "Dance of the Sexes." *Natural History* 114 (55): 34–39.

Querouil, S., Rainer Hutterer, Patrick Barriere, Marc Colyn, Julian C. Kerbis, and Erik Verheyen. 2001. "Phylogeny and Evolution of African Shrews (Mammalia: Soricidae) Inferred from 16s rRNA Sequences." *Molecular Phylogentics and Evolution* 20 (2): 185–195.

Sprawls, S., Kim Howell, and Robert Drewes. 2006. *Reptiles and Amphibians of East Africa*. Princeton, NJ: Princeton University Press.

TLC, Africa. 2008. "Venomous Snakes of Liberia and West Africa." Liberia Telecommunication Corporation. http://www.tlcafrica.com/tlc_snakes.htm.

Webber, W., Lee J. T. White, Amy Vedder, and Lisa Naughton-Treves. 2001. *African Rain Forest Ecology and Conservation: An Interdisciplinary Perspective*. New Haven, CT: Yale University Press.

Welch, K. R. G. 1982. *Herpetology of Africa: A Checklist and Bibliography of the Orders Amphisbaenia, Sauria and Serpentes*. Malabar, FL: Robert Krieger Publishing Company.

Asia-Pacific

Aiken, S. R., and Colin H. Leigh. 1992. *Vanishing Rain Forests: The Ecological Transition in Malaysia*. Oxford: Clarendon Press.

Braatz, Susan.1992. *Conserving Biological Diversity: A Strategy for Protected Areas in the Asia-Pacific Region*. Washington, DC: World Bank.

Brookfield, H. 1995. *In Place of the Forest: Environmental and Socio-economic Transformation in Borneo and the Eastern Malay Peninsula*. New York: United Nations University Press.

Bruenig, E. F. 1996. *Conservation and Management of Tropical Rainforests: An Integrated Approach to Sustainability*. Cambridge: CAB International.

Department of Natural Resources, Environment and the Arts. 2007. "Threatened Species of Northern Australia." Northern Territory Government. http://www.nt.gov.au/nreta/wildlife/animals/threatened/pdf/plants/Schoutenia_ovata_VU.pdf.

Garbutt, N. 2006. *Wild Borneo: The Wildlife, Scenery of Sabah, Sarawak, Brunei and Kalimantan*. Cambridge, MA: The MIT Press.

Jepson, P., J. K. Jarvie, K. MacKinnon, and K. A. Monk. 2001. "The End for Indonesia's Lowland Forests?" *Science* 292: 859–861.

Kira, T., and Keiji Iwata, eds. 1965. *Nature and Life in Southeast Asia*. Kyoto, Japan: Fauna and Flora Research Society.

Knowlton, N. 2001. "The Future of Coral Reefs." *Proceedings of the National Academy of Science* 98 (10): 5419–5425.

Kummer, D. M., and B. L. Turner II. 1994. "The Human Causes of Deforestation in Southeast Asia." *Bioscience* 44 (5): 323–328.

Meijer, W. 1973. "Devastation and Regeneration of Lowland Dipterocarp Forests in Southeast Asia." *Bioscience* 23 (9): 528–533.

Morcombe, M. 2000. *Field Guide to Australian Birds*. Queensland, Australia: Steve Parish Publishing Pty Ltd.

Panayotou, T., and S. Sungsuwan. 1994. "An Econometric Analysis of the Causes of Tropical Deforestation: The Case of Northeast Thailand." In *The Causes of Tropical Deforestation: The Economic and Statistical Analysis of Factors Giving Rise to the Loss of the Tropical Forests,* ed. K. Brown and D. W. Pierce, 192–210. Vancouver: University of British Columbia Press.

Primack, R., and Thomas E. Lovejoy, eds. 1995. *Ecology, Conservation and Management of Southeast Asian Rainforests*. New Haven, CT: Yale University Press.

Robson, C. 2005. *Birds of Southeast Asia*. Princeton, NJ: Princeton University Press.

Rundel, P. W., and K. Boonpragob. 1995. "Dry Forest Ecosystems of Thailand." In *Seasonally Dry Tropical Forests,* ed. S. H. Bullock, H. A. Mooney, and E. Medina, 93–123. Cambridge: Cambridge University Press.

Shuttleworth, C. 1981. *Malaysia's Green and Timeless World*. Singapore: Heinemann Educational Books.

Stott, P. 1988. "The Forest as Phoenix: Towards a Biogeography of Fire in Mainland Southeast Asia." *Geographical Journal* 154: 337–350.

Strahan, R., ed. 1983. *The Australian Museum Complete Book of Australian Mammals*. Sydney, Australia: Angus & Robertson Publishers.

Whitmore, T. C. 1975. *Tropical Rainforests of the Far East*. Oxford: Clarendon Press.

Whitten, T., Sengli J. Damanik, Nazaruddin Hisyam, and Jazanul Anwar. 1999. *The Ecology of Sumatra*. Indonesia: Tuttle Publishing.

Yamada, I. 1997. *Tropical Rain Forests of Southeast Asia: A Forest Ecologist's View*. Honolulu: University of Hawaii Press.

Index

Tropical seasonal forest, 1, 2, 4, 9–11, 141–60; animals, 154; biodiversity, 149; clima-graphs, 145; climate, 144–46; formation and origin, 143–44; geographic location, 142–43; human impact, 156–60; monsoons, 144, 146; soils, 146–48; vegetation, 148–54
Typhoons, 7, 9, 105, 146. *See also* Cyclones; Hurricanes

Understory layers, 3, 50, 184

Vegetation, 2–4, 28–37, 148–54; African rainforest, 79–84; African seasonal forest, 183–88; Asian-Pacific rainforest, 107–12; Asian-Pacific seasonal forest, 199–202; Neotropical rainforest, 50–56;

Neotropical seasonal forest, 164–70; transect, 150; Tropical rainforest, 28–37; Tropical seasonal forest, 148–54

Wallace, Alfred Russell, 106
Wallace's line, 104, 105, 106, 107, 121, 197–98
Wildfires, 74, 159. *See also* Fire
Woody vines, 83, 150, 166; bole climbers, 33–34; danglers, 33, 54; lianas, 33, 36, 83, 165; stranglers, 34, 35, 54

Xenarthra. *See* Edentates

Zambezian subregion, 181, 182, 183, 184. *See also* African tropical seasonal forest

About the Author

BARBARA A. HOLZMAN is professor of geography and environmental studies at San Francisco State University. She has been teaching undergraduate and graduate courses in biogeography and environmental science for more than fifteen years. She has traveled extensively in several tropical regions of the world, particularly Latin America. She is co-author of *Geographical Measurements*, published in 2003 and revised in 2005.

GREENWOOD GUIDES TO
BIOMES OF THE WORLD

Introduction to Biomes
Susan L. Woodward

Tropical Forest Biomes
Barbara A. Holzman

Temperate Forest Biomes
Bernd H. Kuennecke

Grassland Biomes
Susan L. Woodward

Desert Biomes
Joyce A. Quinn

Arctic and Alpine Biomes
Joyce A. Quinn

Freshwater Aquatic Biomes
Richard A. Roth

Marine Biomes
Susan L. Woodward